JIT Implementation Manual

The Complete Guide to
Just-in-Time Manufacturing

Second Edition

Volume 5

JIT Implementation Manual

The Complete Guide to
Just-in-Time Manufacturing

Second Edition

Volume 5

Standardized Operations –
Jidoka and Maintenance/Safety

HIROYUKI HIRANO

CRC Press
Taylor & Francis Group
Boca Raton London New York

CRC Press is an imprint of the
Taylor & Francis Group, an **informa** business

A PRODUCTIVITY PRESS BOOK

Originally published as *Jyasuto in taimu seisan kakumei shido manyuaru* copyright © 1989 by JIT Management Laboratory Company, Ltd., Tokyo, Japan.

English translation copyright © 1990, 2009 Productivity Press.

CRC Press
Taylor & Francis Group
6000 Broken Sound Parkway NW, Suite 300
Boca Raton, FL 33487-2742

© 2009 by Taylor & Francis Group, LLC
CRC Press is an imprint of Taylor & Francis Group, an Informa business

No claim to original U.S. Government works
Printed in the United States of America on acid-free paper
10 9 8 7 6 5 4 3 2 1

International Standard Book Number-13: 978-1-4200-9030-7 (Softcover)

Visit the Taylor & Francis Web site at
http://www.taylorandfrancis.com

and the CRC Press Web site at
http://www.crcpress.com

Contents

Volume 1

1 Production Management and JIT Production Management 1
Approach to Production Management.. 3
Overview of the JIT Production System... 7
Introduction of the JIT Production System.....................................12

2 Destroying Factory Myths: A Revolutionary Approach........... 35
Relations among Sales Price, Cost, and Profit................................35
Ten Arguments against the JIT Production Revolution.......................40
Approach to Production as a Whole 44

Volume 2

3 "Wastology": The Total Elimination of Waste........................145
Why Does Waste Occur?...146
Types of Waste 151
How to Discover Waste 179
How to Remove Waste 198
Secrets for Not Creating Waste..226

4 The "5S" Approach 237
What Are the 5S's? 237
Red Tags and Signboards: Proper Arrangement and
Orderliness Made Visible 265
The Red Tag Strategy for Visual Control 268
The Signboard Strategy: Visual Orderliness 293
Orderliness Applied to Jigs and Tools...307

Volume 3

5 Flow Production **321**

Why Inventory Is Bad 321

What Is Flow Production? 328

Flow Production within and between Factories 332

6 Multi-Process Operations **387**

Multi-Process Operations: A Wellspring for Humanity on the Job.....387

The Difference between Horizontal Multi-Unit Operations and

Vertical Multi-Process Operations 388

Questions and Key Points about Multi-Process Operations.............393

Precautions and Procedures for Developing Multi-Process

Operations 404

7 Labor Cost Reduction **415**

What Is Labor Cost Reduction? 415

Labor Cost Reduction Steps 419

Points for Achieving Labor Cost Reduction 422

Visible Labor Cost Reduction 432

8 *Kanban* **435**

Differences between the *Kanban* System and Conventional Systems...435

Functions and Rules of *Kanban* 440

How to Determine the Variety and Quantity of *Kanban* 442

Administration of *Kanban* 447

9 Visual Control **453**

What Is Visual Control? 453

Case Study: Visual Orderliness (*Seiton*) 459

Standing Signboards 462

Andon: Illuminating Problems in the Factory 464

Production Management Boards: At-a-Glance Supervision.............470

Relationship between Visual Control and *Kaizen* 471

Volume 4

10 Leveling **475**

What Is Level Production? 475

Various Ways to Create Production Schedules 477

Differences between Shish-Kabob Production and Level Production 482

Leveling Techniques 485

Realizing Production Leveling..492

11 Changeover.. **497**

Why Is Changeover Improvement (*Kaizen*) Necessary? 497

What Is Changeover? 498

Procedure for Changeover Improvement 500

Seven Rules for Improving Changeover 532

12 Quality Assurance **541**

Quality Assurance: The Starting Point in Building Products 541

Structures that Help Identify Defects 546

Overall Plan for Achieving Zero Defects.......................................561

The *Poka-Yoke* System 566

Poka-Yoke Case Studies for Various Defects...............................586

How to Use *Poka-Yoke* and Zero Defects Checklists.......................616

Volume 5

13 Standard Operations **623**

Overview of Standard Operations 623

How to Establish Standard Operations 628

How to Make Combination Charts and Standard Operations Charts....630

Standard Operations and Operation Improvements.......................638

How to Preserve Standard Operations...650

14 *Jidoka*: Human Automation... **655**

Steps toward *Jidoka* 655

The Difference between Automation and *Jidoka* 657

The Three Functions of *Jidoka* 658

Separating Workers: Separating Human Work from Machine Work 660

Ways to Prevent Defects 672

Extension of *Jidoka* to the Assembly Line.....................................676

15 Maintenance and Safety **683**

Existing Maintenance Conditions on the Factory Floor.....................683

What Is Maintenance?...684

CCO: Three Lessons in Maintenance 689

Preventing Breakdowns..683

Why Do Injuries Occur?...685

What Is Safety?..688

Strategies for Zero Injuries and Zero Accidents.......................689

Index **I-1**

About the Author...**I-31**

Volume 6

16 JIT Forms **711**

Overall Management 715

Waste-Related Forms 730

5S-Related Forms...747

Engineering-Related Forms 777

JIT Introduction-Related Forms..834

Chapter 13

Standard Operations

Overview of Standard Operations

Why Do We Need Standard Operations?

It so happens that many of the most important elements in the daily activity of manufacturing begin with the letter "M."

In factories, we are trying to find the best possible combination of Men/Women, Materials, and Machines and we develop the most efficient Methods for making things, so that we can make better products while spending less Money.

Standard operations can be defined as an effective combination of workers, materials, and machines for the sake of making high-quality products cheaply, quickly, and safely. As such, standard operations comprise the backbone of JIT production.

Many people make the assumption that standard operations are nothing more than standard operating procedures. But this is not at all the case.

Standard operating procedures have to do with specific standards for individual operations and are just part of what we mean by standard operations. By contrast, standard operations involve the stringing together of individual operations in a specified order to achieve an effective combination for manufacturing products. Another name for standard operations would be "production standards." One might ask why such production standards are necessary in the daily business of manufacturing?

While this may seem like a simple question, it is actually rather difficult to answer. Please think about it for a moment. Why are production standards necessary for daily production activities?

In considering this question, let us suppose that we have asked some other manufacturer to do some manufacturing for us.

The person would probably ask such questions as: "How do you make these products?," "How much time and money does it take to make them?," and "When do you need them delivered?"

Why does the other manufacturer need to know all these things? Basically, because they need to fit the work we have asked them to do into their current production schedule. They will not know whether they can actually make the requested products on schedule unless they have established standard operations. Factories, therefore, need standard operations right from the start.

Standard operations serve the following goals:

1. Quality: "What quality standards must the product meet?"
2. Cost: "Approximately how much should it cost to make the products?"
3. Delivery: "How many products do you need delivered and by when?"
4. Safety: "Is the manufacturing work itself safe?"

At the very least, standard operations should be able to answer those four questions.

It should be clear enough by now why we define standard operations as an effective combination of workers, materials, and machines. We also need to remember that, like all improvement, improvement in standard operations is an endless process.

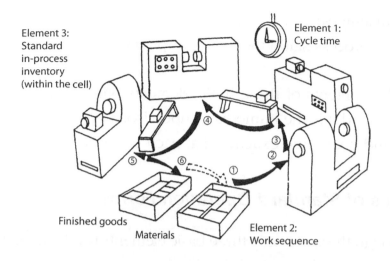

Element 3:
Standard
in-process
inventory
(within the cell)

Element 1:
Cycle time

④

③

②

⑤   ⑥   ①

Finished goods

Materials

Element 2:
Work sequence

Figure 13.1   The Three Basic Elements of Standard Operations.

The Three Basic Elements of Standard Operations

While standard operations involve the effective combination of three "M" elements—men/women, materials, and machines—these elements differ from the three basic elements that go into standard operations. Figure 13.1 illustrates these elements as they are used to create standard operations in a U-shaped manufacturing cell.

Element 1: *Cycle time*

Cycle time is the amount of time it takes a worker to turn out one product (within a cell). We use the production output and the operating time to determine the cycle time.

Element 2: *Work sequence*

This refers to the order in which the worker carries out tasks at various processes as he or she transforms the initial materials into finished goods. It is not the same as the "flow of products" concept we use in flow production.

Element 3: *Standard in-process inventory*

This indicates the minimum amount of in-process inventory (including in-process inventory currently attached to machines) that is required within the manufacturing cell or process station for work to progress.

The contents of these three elements will differ from cell to cell, and it is the immediate supervisor's job to analyze the cell and determine exactly what each element will include.

Types of Standard Operation Forms

Although there are only three basic elements (cycle time, work sequence, and standard in-process inventory) in standard operations, there are five types of standard operation forms.

> ***Form 1****: Parts-production capacity work table*
> This work table examines the current parts-production capacity of each process in the cell. (See Figure 13.2.)
> ***Form 2****: Standard operations combination chart*
> This chart helps us make "transparent" (or obvious) the temporal process of the relationship between human work and machine work. (See Figure 13.3.)
> ***Form 3****: Standard operations pointers chart*
> We use this chart to list important points about the operation of machines, exchanging jigs and tools, processing methods, and so on. (See Figure 13.4.)

Approval stamps	Parts-Production Capacity Work Table		Part No.		Type RY		Entered by Sato	
			Part name 6" pinion		Quantity 1		Creation date 1/17/89	

Process	Process name	Serial No.	Manual operation time (A)		Basic times				Blades and bits		Per unit retooling time F=E+D	Total time per unit G=C+F	Production capacity I/G	Graph time
					Auto feed time (B)		Complet-ion time C=A+B		Retooling amount (D)	Retooling time (E)				Manual work - - - Auto feed ———
			Min.	Sec.	Min.	Sec.	Min.	Sec.						
1	Pick up raw materials	——	1		—		1		——	——	——	1	——	
2	Gear teeth cutting	A01	4		35		39		400	2'10"	0.3"	39.3	717	
3	Gear teeth surface fin.	A02	6		15		21		1,000	2'00"	0.1"	21.1	1,336	
4	Forward gear surface fin.	A03	7		38		45		400	3'00"	0.5"	45.5	619	
5	Reverse gear surface fin.	A04	5		28		33		400	2'30"	0.4"	33.4	844	
6	Pin width measurement	B01	8		5		13		——	——		13	259	
7	Store finished workpiece	——	1		—		1		——	——	——	1	——	

Figure 13.2   Parts-Production Capacity Work Table.

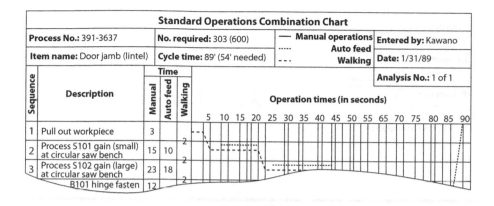

Figure 13.3   Standard Operations Combination Chart.

Summary Table of Standard Operations	Process name		Department	Date	Confirmation		
	Processing sequence						
	Machine number						
No.	Description of operation	Critical factors (correct/incorrect, safety, facilitation, etc.)		Diagram of operation			

Figure 13.4   Standard Operations Pointers Chart.

Work Methods Table	Part no.		Required output	Dept.	Name	Confirmation		
	Part name		Breakdown no.		Date			
No.	Description of operation	Quality		Critical factors (correct/incorrect, safety, facilitation, etc.)	Net time (min. and sec.)	Cycle time	Stand. in-process inv.	Stand. in-process inv. / Safety point / Quality check point
		Check	Measure.					
								● ✚ ◇

Figure 13.5   Work Methods Chart.

Form 4: *Work methods chart*

This chart gives explicit instructions on how to follow standard operations at each process. (See Figure 13.5.)

Form 5: *Standard operations chart*

This chart illustrates and describes the machine layout, cycle time, work sequence, standard in-process inventory,

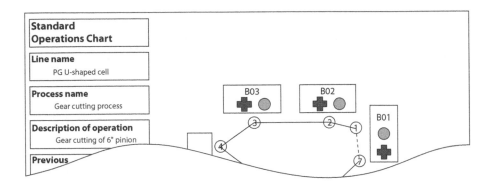

Figure 13.6 Standard Operations Chart.

and other factors in standard operations. Operators should use this chart to check how well they are following standard operations. (See Figure 13.6.)

How to Establish Standard Operations

Transparent Operations and Standard Operations

The first step toward establishing standard operations is to gain a grasp of the way operations are already. To do this, we need to make what is only dimly and vaguely understood as clear and "transparent" (obvious) as possible. This means we have to flush out all of the problems that are hidden within the current situation, look for their causes, and make improvements that will remove those causes and bring about standard operations.

Once we have established standard operations in this way, we still cannot afford to sit back and call the job done. We must repeat the process of flushing out problems and making operations completely transparent. As mentioned earlier, improvement is an endless process. Once we have made improvements, we establish them as standard operations. Then we are ready for another round of problem-hunting to further improve operations and achieve a higher standard. This spiral of improvement in standard operations is illustrated in Figure 13.7.

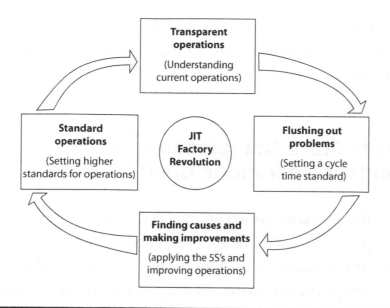

Figure 13.7   Spiral of Improvement in Standard Operations.

Steps in Establishing Standard Operations

Establishing standard operations is a four-step process, as described below.

> ***Step 1***: *Create a parts-production capacity work table*
> List the processing capacity of each cell or process station as it currently stands.
>
> ***Step 2***: *Create a standard operations combination chart*
> Time manual operations, auto feed operations, and walking to elucidate the relationship between human work and machine work.
>
> ***Step 3***: *Create a work methods chart*
> The workshop will need one of these for passing along instructions to new workers.
>
> ***Step 4***: *Create a standard operations chart*
> This schematic chart will provide a visual aid for quickly learning the machine layout, work sequence, and other important factors.

That is all there is to it. Usually, we can incorporate the standard operations combination chart with a standard operations chart to provide a useful reference chart for posting on the wall in the workshop. Figure 13.8 shows an example of such a combined chart.

How to Make Combination Charts and Standard Operations Charts

Even after we have gained an intellectual grasp of what standard operations combination charts and standard operations charts are all about, it is not always easy to actually create one. Perhaps the following exercise can serve as a reference for those who are about to attempt establishing standard operations for the first time in their workshops.

Exercise in Making Combination Charts and Standard Operations Charts

Using the parts-production capacity work table shown in Figure 13.9, make a combination chart and standard operations chart to suit the following two conditions:

Condition 1: Work sequence of processing—Raw materials →A01→A02→A03→A04→B01→finished goods

Condition 2: Required output is 613 units per day

1. Take 7 hours and 50 minutes as the amount of time per worker day, with no short breaks.
2. Take 2 seconds as the walking time for every instance of walking.
3. To keep this exercise simple, do not calculate change-over time.

Steps in creating charts:

1. Calculate the cycle time. To obtain the cycle time, divide the operating time per day by the required output per day.

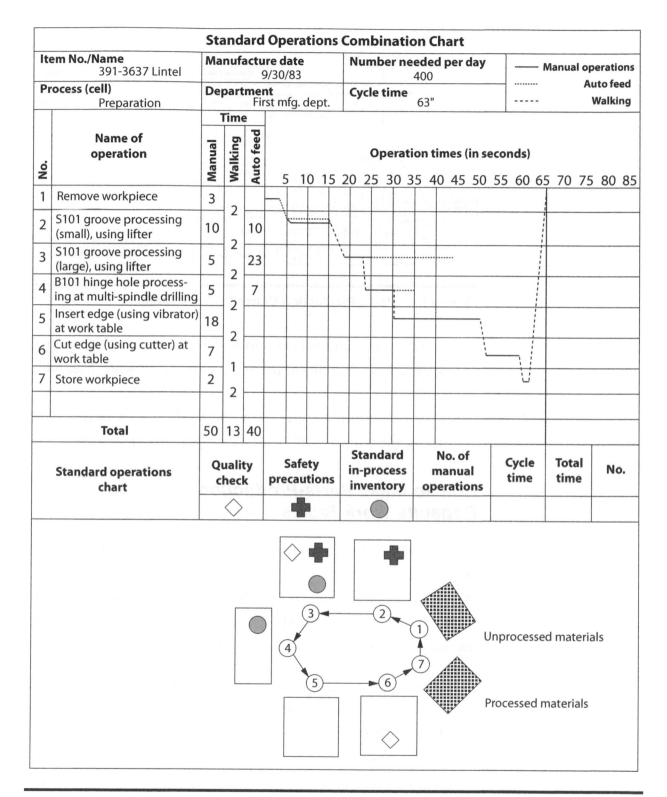

Figure 13.8 Standard Operations Combination Chart with Standard Operations Chart (Schematic).

Approval stamps	**Parts-Production Capacity Work Table**		Part No.		Type *RY*		Entered by *Sato*	
			Part name *6" pinion*		Quantity *1*		Creation date *1/17/89*	

Process	Process name	Serial No.	Manual operation time (A) Min. Sec.	Basic times — Auto feed time (B) Min. Sec.	Complet-ion time C=A+B Min. Sec.	Blades and bits — Retooling amount (D)	Retooling time (E)	Per unit retooling time F=E+D	Total time per unit G=C+F	Production capacity I/G	Graph time — Manual work - - - / Auto feed ———
1	Pick up raw materials	—	1	—	1	—	—	—	1	—	
2	Gear teeth cutting	A01	4	35	39	400	2'10"	0.3"	39.3	717	4" - - 35" - - -
3	Gear teeth surface fin.	A02	6	15	21	1,000	2'00"	0.1"	21.1	1,336	6" - 15"
4	Forward gear surface fin.	A03	7	38	45	400	3'00"	0.5"	45.5	619	7" - - 38" - - -
5	Reverse gear surface fin.	A04	5	28	33	400	2'30"	0.4"	33.4	844	5" - 28" - -
6	Pin width measurement	B01	8	5	13	—	—		13	259	8" 5"
7	Store finished workpiece	—	1	—	1	—	—	—	1	—	
	Total		32	2 01	2 33	Daily operating time (I): 7 hours, 50 minutes				28,200 seconds	

Figure 13.9   Parts-Production Capacity Work Table.

2. Create the standard operations combination chart. Drop a thick red line along the time axis to indicate the cycle time.

3. Create a standard operation chart. The point of this is to show the amount of standard in-process inventory.

How to Make Parts-Production Capacity Work Tables

Figure 13.9 shows the parts-production capacity work table to be used in the above exercise. The following shows how the standard operations combination chart and standard operations chart should look when completed. First, the following are steps for filling out these charts:

1. Assign sequential numbers to indicate the work sequence.
2. Enter the process name.
3. Enter the machine's serial number.
4. Basic times:
   a. Manual operation time (_____): Enter the time required by the worker to perform each operation in the cell.
   b. Auto feed time (_____): Enter the amount of "machine work" time.

c. Completion time: Enter the amount of time required for one workpiece to be completed (from start to finish in the cell).

Completion time = Manual operation time + auto feed time (if operations are performed serially)

5. Blades and drill bits.

a. Retooling volume: Enter the number of blades or bits to be exchanged.

b. Retooling time: Enter the total time required for retooling.

6. Per-unit time = Completion time + per-unit retooling time

7. Production capacity: Enter the number of units that can be produced in one standard day (= daily operating time/per-unit time).

8. Graph time: Enter the operating time (_____) and the auto feed time (_____) onto a graph. For example, for work sequence Step 2, enter the two lines as shown below to provide an easy-to-grasp indication to use when creating a standard operations combination chart.

Three patterns for the standard time are as follows:

Pattern 1: Serial Operations

In this case, the machines' auto feed operations begin only after the worker's manual operations end. Thus, the two follow each other in a series with no overlap (that is, human work and machine work are completely separate), as follows:

Pattern 2: Partially Parallel Operations

Here, the machine begins its work while the worker is still busy. The worker begins before the machine joins in and the machine keeps operating after the worker has finished.

This still allows some room for the separation of human work and machine work. The overlap between the two should be indicated as follows:

Pattern 3: Parallel Operations

In this case, the machine is completely unable to operate without human assistance, and thus there is no separation between human work and machine work, as is demonstrated in the following example.

How to Make Standard Operations Combination Charts

Figure 13.10 shows a standard operations combination chart that was filled out from the above exercise. If you wish to perform the exercise and complete your own standard operations combination chart, please compare it afterward with the one in the figure.

The steps for filling out the standard operations combination chart are described below.

> ***Step 1****: Draw a red line to indicate the cycle time.*
> Cycle time = Total operating time/required output
> ***Step 2****: Calculate whether the cell can be handled by just one worker.*
> Using the parts-production capacity work table from the above exercise, see whether or not the sum of the manual working time and the walking time is less than the cycle time.
> ***Step 3****: Enter a description of the process operations under the "Description of Operations" column.*

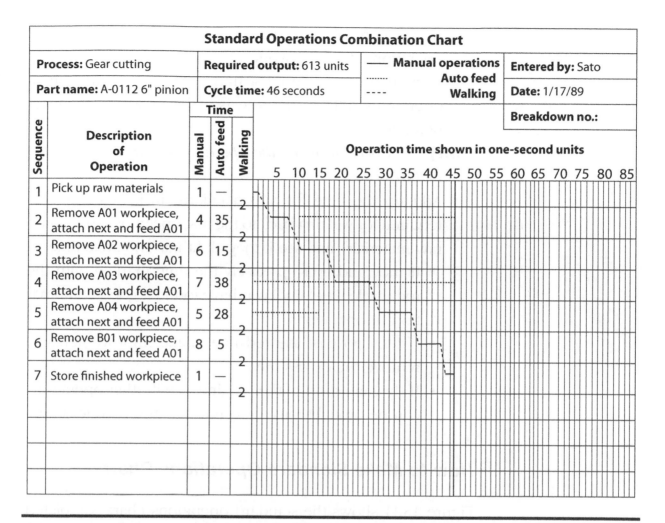

Figure 13.10   An Example of a Standard Operations Combination Chart.

Step 4*: Enter the various time measurements under the "Time" column.*

Step 5*: On the graph, draw solid lines for manual work time, broken lines for auto feed time, and wavy lines for walking time.*

If the auto feed time exceeds the cycle time, enter the extra time from the zero (start) position in the graph.

Step 6*: Check the combination of operations.*

When the auto feed time exceeds the cycle time and some of it must be entered from the zero position, it may overlap with the manual operation time. If it does, it indicates the manual work must wait for the auto feed (machine work) to finish, which means that the combination of operations does not work.

In such cases, we must find a better combination of operations. Idle time waste is to be avoided whenever possible.

Step 7: *Check whether the operations can be completed within the cycle time.*

Add up the time for all operations, including the time required for walking back to the first operation (picking up raw materials), and see if they all fit into the cycle time.

- If they add up to precisely the time marked with the red (cycle time) line, you have found a good combination of operations.
- If they go past the red line, make improvements to remove the excess time.
- If they fall short of the red line, see if other operations can be brought into the cell to reach the cycle time.

How to Make Standard Operations Charts

Figure 13.11 shows the standard operations chart completed from the exercise described in the previous section. After making your own standard operations chart, be sure to compare it to this one.

The following are the steps for filling in the standard operations chart.

Step 1: *Enter the work sequence.*

Enter circled numbers next to the machines to indicate the order in which they are used during the work sequence, then connect the machines with a solid line, as shown in Figure 13.11. Draw a broken line between the last step and first step in the work sequence.

Step 2: *Enter the quality check points.*

Enter diamond symbol next to all machines that require quality checks.

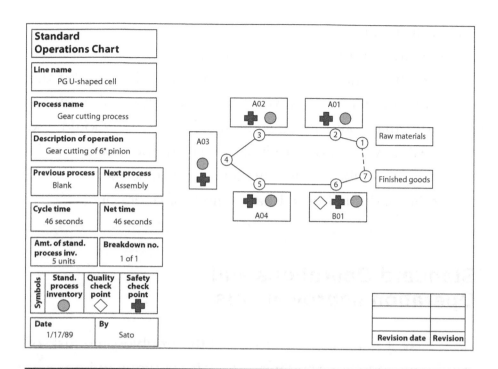

Figure 13.11   Standard Operations Chart.

Step 3*: Enter the safety check points.*

Enter cross symbols next to all machines that require safety checks. Be sure to enter one of these marks next to any machine that uses a blade.

Step 4*: Enter the symbols for standard in-process inventory.*

Enter shaded circle symbols where standard in-process inventory is required for whatever reason (separating human work and machine work, balancing processes, and so forth).

Step 5: *Enter the cycle time.*

Step 6: *Enter the net time.*

Enter the operation time for the sequence shown in the circled numbers. Do not include the time taken up by quality checks or blade exchanges that are done less than once per cycle.

Step 7*: Enter the amount of in-process inventory.*

In this box, enter the number of shaded circles you entered in the graph at Step 4. Separation during auto feed counts as one unit of in-process inventory.

Step 8*: Enter the breakdown number.*

Usually, both the standard operations combination chart and the standard operations chart are filled out by the same operator. However, sometimes the cell requires more than one operator, in which case we should use breakdown numbers to indicate which operator is which.

- First number = Operator's number in sequence
- Second number = Total number of operators

Standard Operations and Operation Improvements

How easy it is for factories to avoid the troublesome task of improving operations and instead opt for equipment improvements. One of the purposes of improvement is to lower costs, but companies find themselves spending a fortune on new or remodeled equipment, all in the name of making improvements. A factory's choice of equipment should be based on the needs of production operations, but many factories put the cart before the horse by changing production operations to suit the equipment. Production machines are tools for production and it makes no sense to have production suit the tools rather than vice-versa.

The following are a few examples of what we mean by "operation improvements."

Improvements in Devices That Facilitate the Flow of Goods and Materials

There are basically two ways to change the devices that facilitate the flow of goods and materials. One is to bring equipment closer to each other in the cell and arrange them according to the work sequence. This creates a "flow shop" type of workshop and is known as "layout improvement."

The other way is to switch from large-lot processing to small-lot or one-piece flow. This is called "flow unit improvement." Each of these types of improvement should initially be used to remove major forms of waste.

Improvement from Specialized Operations to Multi-Process Operations

Conventionally, factories assigned very specialized tasks to each worker, and workers generally remained at one place to perform those tasks while the in-process inventory was conveyed by hand or conveyor belt. This system required workers to spend a lot of time going to pickup things or put things down. We can eliminate the waste inherent in such specialized operations by training workers in the multiple skills needed to conduct multi-process operations, in which a single worker guides each workpiece throughout all of the workshop's processes with a minimum of walking waste.

Improvement of Motion in Operations

Whenever a worker takes a step or stretches out an arm, "motion waste" is created. Conventional industrial engineering has developed a method of motion analysis to identify wasteful motion. Wasteful motion can be caused by a poor equipment layout or sloppy housekeeping of parts and tools. We must reduce this kind of waste by making the equipment layout and organization of parts and tools more conducive to efficient operations.

Improvement by Establishing Rules for Operations

Operational procedures cannot be readily understood and followed by new workers if they vary from one worker to the next. It is only when the correct operational procedures have been clearly established as strictly enforced rules that everyone will perform operations the same way. Along with rules for correct procedures, there must also be rules that help establish level production.

Once we have laid the groundwork by improving operations, we are ready to begin thinking about how the equipment might be improved to better suit the improved operations. The following are a few ways to improve equipment.

Improve the Equipment to Better Serve Operations

Quite often, improved operations do away with a prior need for large equipment that can handle large lots or operate at high speed. Instead, the improved operations tend to call for smaller, slower, and more specialized equipment that can be counted on to produce high quality and be brought directly into the processing or assembly line.

Make the Machines More Independent to Separate People from Them

If the operator must press a switch and then hold the workpiece in place while the machine processes it, we should remodel the machine so that it can operate without human assistance or supervision. In JIT, this is called "separating people from machines," and it allows people and machines to work independently to add value to products simultaneously.

Improving Equipment to Prevent Defects

We can equip machines with detectors and switches that enable the machine to automatically detect defects (or potential defects), stop operating, and issue an alarm. Such devices are a key means of preventing defects.

It bears repeating that operation improvements should be made before equipment improvements. It should also emphasize that the most effective means of removing motion-related waste from operations is to make "operational device improvements." This means first changing the flow unit from large lots to small lots or one-piece flow, then changing the equipment to suit the new flow method.

Improving the Flow of Materials

The most important kind of operation improvement we can make is to change the way goods flow through the factory. However, such a change is not possible unless we are willing to give up the way we have been doing things and undergo an "awareness revolution" that negates the old tried-and-true methods as the worst possible methods.

In other words, changing the flow of goods requires changing our way of thinking, all our concepts about equipment and how to arrange it, and, most importantly, our ideas about how goods should proceed through the production line. We need to change just about everything that goes on in the factory.

Figure 13.12 shows an example of how the flow of goods was improved at a solder printing process for semiconductor wafers.

Before improvement, this processing line was run by four operators, each of whom worked independently of the other three. The line operated in 600-unit batches and used a large dryer. Sending such large lots through was a start-and-stop operation that reflected precious little ingenuity and resulted in frequent bottlenecks.

The improvement included training a single operator in the skills needed to handle five processes: printing (the front of the wafer), baking, printing (the back of the wafer), input to the reflow oven, and output from the reflow oven. The layout was changed to facilitate these tasks and to minimize motion-related waste. The reflow jig was changed to accommodate "two-piece" flow. They got rid of the large dryer, brought a compact ultraviolet-ray dryer out of storage and remodeled it to serve in place of the large dryer, but in an "in-line" location. Finally, they attached a return conveyor at the back of the reflow oven to match up the oven's input and output sites. As a result, they were able to cut the required manpower in half while doubling productivity.

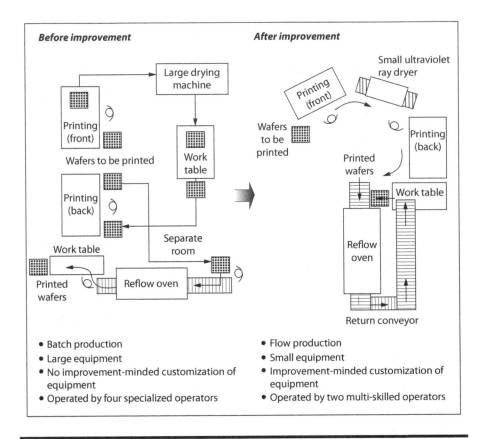

Figure 13.12   Improved Flow of Goods at a Solder Printing Process for Semiconductor Wafers.

Improving the Efficiency of Movement in Operations

Not all of what factory workers do on the job can truly be called "work" in the sense of adding value to goods. On the contrary, most of what the typical factory worker does adds no value. It is therefore not work, just motion. Motion study is an industrial engineering technique that helps distinguish between productive work and nonproductive motion in order to raise the work-versus-motion ratio.

When we use motion study to remove wasteful motion from operations, we try to make the job easier, and with more economical movement, more efficient work sequences, and better combinations of tasks.

The "principles of economy of motion" can be a very good tool for improving the motions of workers to remove

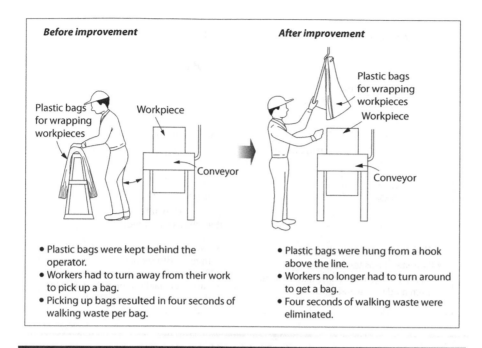

Before improvement

After improvement

Plastic bags for wrapping workpieces

Workpiece

Plastic bags for wrapping workpieces

Workpiece

Conveyor

Conveyor

- Plastic bags were kept behind the operator.
- Workers had to turn away from their work to pick up a bag.
- Picking up bags resulted in four seconds of walking waste per bag.

- Plastic bags were hung from a hook above the line.
- Workers no longer had to turn around to get a bag.
- Four seconds of walking waste were eliminated.

Figure 13.13   Improvement in Placement of Parts.

waste from human actions. (For further description of the "principles of economy of motion," see Chapter 3). Following these principles helps "tighten the cost belt" by removing the "fat" in the form of the 3 *Mu*'s (*muda* or waste, *mura* or inconsistency, and *muri* or irrationality). Naturally, this means improving human movements, but it also involves improvements in the ways thing are placed, the arrangement and use of jigs and tools, and the organization of the entire work environment.

1. Improvement in Placement of Parts

Figure 13.13 shows one improvement that involved moving a set of plastic bags used for wrapping workpieces from behind the operator and hanging them from a hook above the line to keep them within easy reach. This simple improvement saved four seconds of walking waste (per unit).

2. Improvement in Picking Up Parts

Figure 13.14 shows an example of how picking up parts at an assembly line was improved. Before the improvement, the

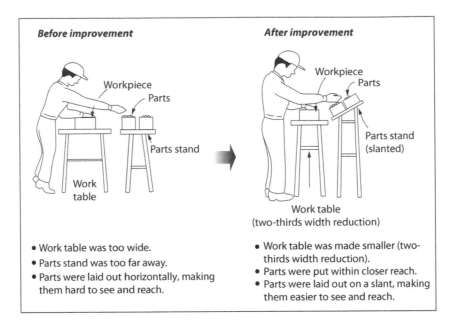

Figure 13.14   Improvement in Picking Up Parts.

parts were kept on a large work table located a little too far from the assembly line. All of the parts were laid out on the same horizontal level, making them hard to see and reach.

As part of the improvement, the work table was reduced to the minimum required size, was moved closer to the assembly line, and the parts boxes were set-up on a higher, slanted stand to make seeing and reaching them easier.

3. Improvement from One-Handed Task to Two-Handed Task

Figure 13.15 shows how the task of assembling push buttons on telephones was improved from being a one-handed task to a two-handed task.

Before the improvement, there was no jig to hold the workpiece in place. Instead, the assembly worker had to hold down the workpiece with her left hand while using her right hand to insert the push buttons one by one.

After the improvement, the assembly worker simply sets the workpiece into a stabilizing jig and then can use both

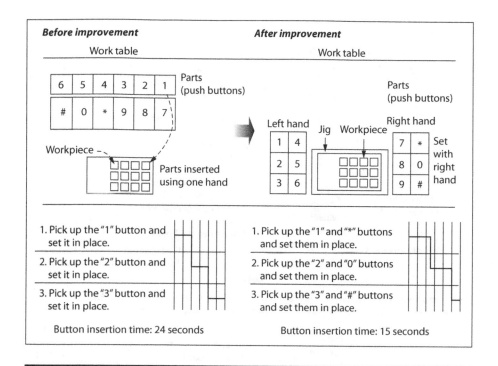

Figure 13.15   Improvement from One-Handed Task to Two-Handed Task.

hands to insert the push buttons. In addition, the arrangement of push buttons to be inserted was changed to match their arrangement after insertion. This helped to keep workers from accidentally inserting push buttons in the wrong places.

4. Improvement That Eliminates Walking Waste

Figure 13.16 shows an improvement example in which walking waste was removed from speaker cabinet processing operations.

This workshop had been using the conventional layout in which each machine was operated by a different worker, each of whom picked up workpieces from large piles of in-process inventory. Obviously, such a layout is not conducive to the concept of cycle time, and instead they tried to maintain a 33-second pitch, beginning at the process where V cuts were made in the speaker cabinets' processed particle boards.

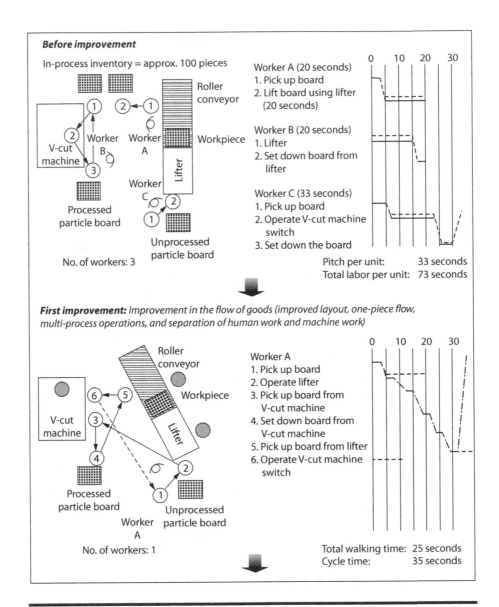

Figure 13.16 First and Second Improvements in Speaker Cabinet Processing Operations.

■ The workshop was run by three workers.

■ There were about 100 pieces of in-process inventory.

■ The pitch per unit was 33 seconds.

■ The total labor per unit was 73 seconds.

As a first improvement, a fundamental change was made in the flow of goods. The V-cut machine was installed in a pit and could not be moved, so they moved the lifter as close to the V-cut machine as possible. Once before, the lifter had

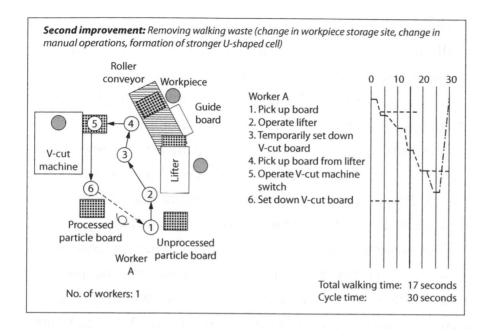

Figure 13.16 (continued)

been moved closer to the V-cut machine, but this was not understood as an improvement at the time. The distance the lifter could be moved was restricted by the electrical cord, and no extension cord was available in the factory. Therefore, they had to compromise in improving the layout.

In the first improvement, they managed to reduce the labor force from three workers to just one by establishing multi-process operations. Naturally, this change included eliminating the stack of in-process inventory between the lifter and the V-cut machine. Fortunately, worker A (the single remaining worker) was an old hand in that factory who was able to pickup the "one piece flow" way of doing things quite readily. Both the lifter and the V-cut machine could feed the workpieces downstream automatically, which enabled the separation of human work and machine work. These changes brought the following results:

- Reduction of labor force from three workers to one.
- Reduction of total in-process inventory to just three workpieces.
- Establishment of a 35-second cycle time.

The improvement, however, was not totally satisfactory. First of all, the worker had to walk a rather complicated pattern to complete the work cycle. Whenever we have complexity, we usually have waste, and it pays to remember "simple is best." Improvement team members counted 25 steps taken by the worker during the work cycle, which means 25 seconds of walking waste (each step is roughly equal to one second of waste). These drawbacks led improvement team members to regroup and launch a second improvement effort.

They determined that they needed to make the equipment layout more compact, but they were faced with the problem of the lifter's fully extended power cord which prevented them from moving the lifter any closer to the V-cut machine. The roller conveyor had no power cord and could be moved freely, although they ended up "bending" the roller conveyor so that its output end is close to the V-cut machine, as shown at the bottom of Figure 13.16.

They then wondered if the roller conveyor could convey the particle boards at its new angle without dropping them. They tried one board; the conveyor dropped it and ruined it.

Then they started brainstorming for solutions to this problem. They tried attaching a guide board to the side of the roller conveyor to keep the particle board from dropping. It worked.

Next, they found a way to avoid having to move the boards in a direction opposite that of the processing flow. To do this, they established a temporary storage site for boards output from the V-cut machine and changed the work sequence around, as shown at the bottom of Figure 13.16. This reduced walking time, which was 17 seconds after the first improvement, to just eight seconds. It also resulted in a five-second reduction in the cycle time, going from 35 seconds after the first improvement to 30 seconds.

If we compare the results of the second improvement to the way things were before the first improvement, we can note the following:

- Workforce reduced to one (reduction of two workers).
- In-process inventory reduced to four workpieces (reduction of about 96 workpieces).
- Pitch per unit (cycle time) reduced to 30 second (reduction of three seconds).
- Total labor per unit reduced to 30 seconds (reduction of 43 seconds).

Both the first and second improvements were made right away, before people had time to apply for money for expensive improvements. The two improvements cost nothing but realized dramatic cost savings. They estimated that the cost savings were roughly proportional to the time invested in studying means of improvement.

Improving the Separation of Worker

Figure 13.17 shows how an improvement involving separation of human work and machine work was achieved for a groove processing operation that uses a lifter.

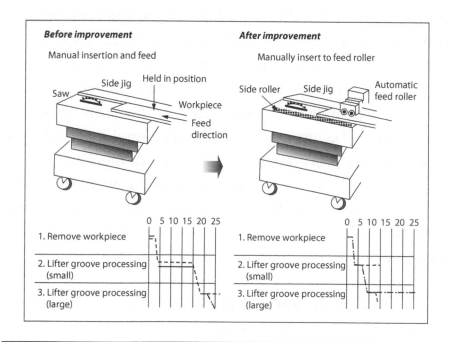

Figure 13.17 Separation of Human Work and Machine Work at a Groove Processing Lifter.

Before the improvement, the operator had to use both hands to align the workpiece along the side jig on top of the lifter and then had to push the workpiece along as the groove was cut. This meant that the operator was unable to separate himself from the machine at any time during the process.

The improvement included attaching a roller to the top of the lifter so that workpieces could be fed automatically over the groove cutter and a side roller to keep the workpiece from shifting sideways. These devices allowed the operator to separate himself from the machine once he had set the workpiece against the rollers and shortened the groove processing cycle time by eight seconds, as shown in Figure 13.17.

How to Preserve Standard Operations

Standard Operations and Multi-Skilled Workers

Once we have established standard operations, it is by no means a given that the workshop's operators will be able to perform them right away. It takes time to get used to the new procedures and to become proficient in them. Usually, each operator works a little differently, and the first task is to eliminate such individual differences. At this point, it is vital that operators be given a lot of guidance until they feel they know the new procedures like the backs of their hands.

We must be extra careful when training workers in the multiple skills they will need for multi-process operations. Workers should gradually expand the range of their skills, and not go any faster than they are able in learning new ones.

Figure 13.18 shows how a U-shaped manufacturing cell was used for on-the-job multiple skills training for operators. In the figure, the trainee (worker A) is able to perform only the first five steps before the cycle time is up, then returns to

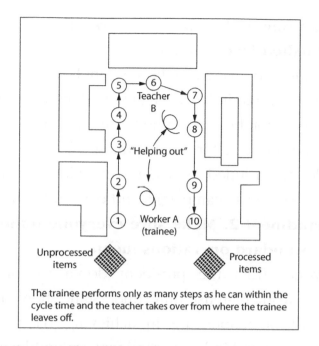

The trainee performs only as many steps as he can within the cycle time and the teacher takes over from where the trainee leaves off.

Figure 13.18   Multiple Skills Training.

Step 1. At Step 6, the teacher takes over and performs the rest of the steps in the work sequence.

Gradually, the trainee is able to take on additional steps and still remain within the cycle time. The trainee may perform Steps 1 to 7 for a while, then move on to Steps 1 to 8, 1 to 9, and finally the entire 10-step process.

The Ten Commandments for Preserving Standard Operations

I loathe to recall how often I have seen people work hard to establish rules for standard operations, only to stash the rules away in some desk drawer and forget about them. It makes me wonder why they even bothered to make the rules in the first place.

Please remember that standard operations are meaningless unless they are maintained.

The following are "ten commandments" that have evolved over the years for preserving standard operations.

Commandment 1: Standard operations must be established factory-wide.

No matter how often or how strongly the factory-floor workers are reminded to maintain standard operations, they will not be maintained unless top management gets behind the effort. Maintaining standard operations should be included as a company-wide project, along with zero-defects campaigns and cost-cutting activities.

Commandment 2: Make sure everyone understands what standard operations mean.

Everyone—from the president down to the newest factory worker—must fully understand how important standard operations are in achieving JIT production. Study group and in-house seminars are good ways to get this message across.

Commandment 3: Workshop leaders must be confident in their skills when training others in standard operations.

Training workers in the new procedures called for by standard operations will go much more smoothly when the workshop leaders who do the training are positive and confident about the change to standard operations. The leaders should appear as if they had already been making things the new way for years.

Commandment 4: Post reminders in the workshop.

Once standard operations have been established at a workshop, signboards and other visual tools should be used to remind workers of their duty to maintain the standard operations.

Commandment 5: Post standard operations signs in obvious places.

Post signs containing graphics- and text-based descriptions of the standard operations at places where the workers can see them easily and compare their own operational procedures to those described on the signs.

Commandment 6: When necessary, get a third person to help out.

Sometimes, bringing in a well-trained new person from some other department is a good way to clear up misunderstandings in learning and maintaining standard operations.

Commandment 7: Reprimand the workshop leader when standard operations are not being maintained.

When workers' actions or work sequences differ from those prescribed by standard operations, we have proof that standard operations are not being maintained. When a factory manager discovers this, instead of chewing out the workers, he should reprimand the workshop leader, right there in front of everyone. This tactic is more effective in strengthening the bond between workshop leaders and their charges.

Commandment 8: Reject the status quo.

Improvement is endless. Even after standard operations have been established, workshop operators cannot afford to become complacent in the belief that they have found the optimum method of operations. It is much better if they believe that the status quo—no matter how successful—is a bad system that must be improved. Only then will their minds remain open to the possibility of further improvement.

Commandment 9: Conduct periodic improvement study groups.

Improvements must be carried out continually. The longer improvements continue, the stronger the company becomes. Unless we work to improve things, they tend to backslide. Strong manufacturing companies are ones that "keep the ball rolling" by sponsoring regular improvement study groups to review current conditions and study possible improvements.

Commandment 10: Take on the challenge of establishing new standard operations.

There is always room for improvement. To establish a new and better set of standard operations, we need to take another critical look at current conditions, flush out the inherent problems, and implement improvements.

The place to discover needs for improvement is in the workshop. Just stand there and watch closely for five minutes. Odds are that the workshop will show you several things in need of improvement. You do not have to think them up—they just come naturally.

Chapter 14

Jidoka
Human Automation

Steps toward *Jidoka*

There are many ways to make the same product. Sometimes all it takes is a very simple tool to process the workpiece. Other times, workers are using both hands to hold something in place during processing when a simple jig could do the trick just as well. Sometimes we can let the machine do part of the work and sometimes we can let the machine do all of it. In other words, there are many ways—various operational methods and flow methods—we can use to make similar products.

There are four steps we should take in developing *jidoka,* and each of these steps is concerned with the relationship between people and machines.

Step 1: Manual labor

Manual labor simply means that all of the work is being done by hand. This makes sense only when the labor costs are cheap and/or the manual work can be done very quickly, such as in the manual assembly line shown in the photograph.

Step 2: Mechanization

Mechanization means leaving part of the manual operations to a machine. We have reached a stage where the work is shared between the worker and the machine, but the worker still does the lion's share. (See photo.)

Step 3: Automation

At this step, all manual labor in processing is taken over by the machine. The worker just sets the workpiece up at the machine and presses a switch to start the machine. The worker can leave the machine alone at that point, but there is no way to know whether the machine is producing defective goods. (See photo.)

Step 4: *Jidoka* (human automation)

As at the automation step, the worker simply sets up the workpieces, presses the ON switch, and leaves the machine to do the processing. In this case, however, the worker need not worry about defects. The machine itself will detect when a defect has occurred and will automatically shut itself off. In addition to defect detection devices, *jidoka* sometimes includes auto-input (auto-feed) and auto-output (auto-extract) devices that completely eliminate the need for worker participation.

The Difference between Automation and *Jidoka*

In an earlier chapter, we discussed the distinction between "moving" and "working" as it pertains to workers' on-the-job activities. The same thing can be said about machines: Sometimes machines are actually working (adding value to something), and at other times they are just moving. How many factories have introduced expensive new machinery to automate and cut labor costs only to discover that, once the machines are operating, there are suddenly new demands for human labor? Perhaps a certain machine cannot do the entire job as planned and requires some human assistance. Or maybe another machine tends to spurt out defective goods and requires a human supervisor. When they add up all the costs, it turns out that they are losing money by automating.

The reason for this all-too-common problem is that the machines are allowed to "move" instead of "work." Or rather, people think that as long as the machines are moving, they are working. But what good does automation equipment do if it cannot actually handle the entire process or if it keeps running even when it produces defective goods? Eventually, such machines need a human supervisor.

By contrast, *jidoka* enables factories to keep equipment running without human assistance or supervision. Current equipment can be upgraded cheaply as "human automated" machines, which actually work while they move and do not disrupt the flow of goods. Indeed, were it not such a mouthful, we might well call them "flow-oriented human automated machines."

Separating workers from machines is not a one-step process. First, we must analyze the worker's operations, then apply *jidoka* to each of them, one at a time. Bold schemes to fully automate in one fell swoop always end up costing a fortune. And, interestingly enough, the more money we spend in automating, the more the new equipment is likely to disrupt the flow of goods. Instead, we need to keep in mind the ratio of labor costs to equipment costs at each step of the way. That is why *jidoka* must proceed carefully, one step at a time.

The Three Functions of *Jidoka*

Jidoka starts by looking at operations that are being performed manually or only partially by machine, distinguishing the human work from the machine work, then taking a closer look at the human work. During each part of the manual operations, we need to ask, "What is the worker's right hand doing?," "What is his left hand doing?," and so on. Then we can ask, "How can we free his left hand from having to do something?" and "How can we free his right hand?" Gradually, we reduce the human work and increase the machine work.

It makes sense to mechanize or automate when the result is lower costs and higher productivity, such as when using an electric motor frees the left hand or using some mechanism frees the right hand. Freed hands can be used for other work. Once we have gotten to the point where the worker's hands and feet are all free after the machine starts operating, we can physically separate the worker from the machine. In JIT, we call this separating human work from machine work. However, as mentioned earlier, it does no good to separate people from machines if the machines cannot be trusted to continue producing high-quality products. Neither does it save money to have the machine do the work while a worker stands by watching out for defects. After all, the whole point of automation is to cut costs.

So, the key is to develop automated machines that do not produce defective goods. To do that, we have to apply human wisdom to change machines that merely "move" into ones that "work." The development of defect-prevention devices for automated equipment is the heart and soul of *jidoka*. The machines must be able to detect by themselves when defects occur, stop themselves, and sound an alarm to inform people about the abnormality. The machine does not have to be able to tell what kind of abnormality has occurred— especially since abnormalities vary widely among different machines, processes, and users—but they do need to let the nearby people know that something strange has happened. The companies that make the manufacturing equipment do not know exactly how their equipment will be used; it is up to the users to customize it to suit their particular needs.

When we have customized our manufacturing equipment to operate reliably and automatically without the risk of turning out an endless stream of defective goods, a single worker can handle several machines or even several groups of machines. Imagine how high productivity soars when that happens!

We usually start by applying *jidoka* to processing equipment. If we succeed at that, we are ready to take on the challenge of bringing *jidoka* to assembly operations. On assembly lines, the purpose of *jidoka* is to get operators to press the stop button (the red "emergency" button) whenever any kind of defect, missing part, omitted task, or other abnormality occurs. Once they have stopped the line this way, they need to make an immediate improvement to solve the problem. They also need to constantly strive to eliminate various forms of waste from their operations to keep raising productivity.

The three main functions of *jidoka* can be summarized as follows:

Function 1: Separation of human work from machine work. *Jidoka* calls for the gradual shifting of all human work to machine work, thereby separating people from the machines.

Function 2: Development of defect-prevention devices. Instead of requiring human supervisors, machines should have the ability to detect and prevent the production of defective goods. Such machines are truly "working" and not just "moving."

Function 3: Application of jidoka *to assembly operations.* Like processing equipment, assembly lines must be stopped as soon as a defect occurs and corrective measures must be taken right away.

Separating Workers: Separating Human Work from Machine Work

What Does Separating Workers Mean?

I remember a factory visit during which one of the company's top managers took special pains to point out a recent acquisition—a late-model numerically controlled machining center. Full of pride, he had us watch the new machine at work. An operator pushed the start button and then stood by throughout the entire two-minute process, just keeping an eye on what was happening.

Naturally, I asked the manager why the operator was staying by the machine. The manager pointed out several reasons—the machine spurts out metal shavings, the operator needs to make sure it is operating correctly, and so on. In other words, the operator had merely switched jobs. Instead of being an operator, he was now a supervisor.

So there it was, the latest in NC machine technology, and still worthless as far as cutting costs goes. I suppose its greatest value to the company was as an amusing new "toy" for the top managers to show off to visitors—evidence that the company was keeping up with the latest fashions in modernization. No one seemed to be paying any attention to what the new machine meant in terms of improving the production system.

Consider, for example, the production configuration shown in Figure 14.1. There are three operators (A, B, and C), each of whom is assigned to one of three machines (1, 2, and 3). After the operators finish their manual task, they set the workpiece

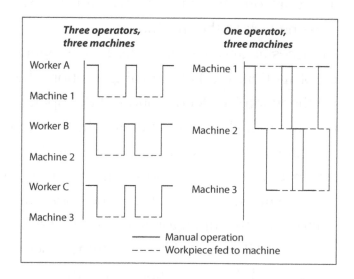

Figure 14.1   Separating Workers from Machines.

into the machine and wait for the machine to go through its process, thereby creating idle time waste.

To remove this idle time waste, the company decided to implement *jidoka*. First, they remodeled the machines to separate the workers from them. Next, they changed the equipment layout to bring the machines closer together. This made it possible for just one worker to handle all three machines consecutively, eliminating idle time waste. The key improvement that made this productivity-boosting overhaul possible was separating workers; that is, separating human work from machine work.

Procedure for Separating Workers

What is the best way to go about separating workers from their machines? For example, if part of Worker X's job is to use his left hand to hold a workpiece against a drilling machine while the machine drills holes into the workpiece, how can he separate himself from the drilling machine? Let us also suppose his job includes using his right hand to turn a wheel that feeds workpieces into a lathe. How on earth can he leave the machines to do all the work? That is precisely what we need to figure out. We must enable him to leave every single processing task to the machines.

Consider lathes as another example. Lathes operate using three kinds of motion: the lathe turning motion, the cutting motion, and the workpiece feed motion. If the operator needs to assist the lathe in making any of these kinds of motion, he cannot be separated from the lathe. (See Figure 14.2.)

If, for instance, the operator's job consisted only of guiding the bite's lateral motion and the lathe took care of the two other motions, the operator still cannot be separated from the machine. Likewise with the drilling machine mentioned above, the drilling machine will often execute the drill's rotary motion and the workpiece feed motion while

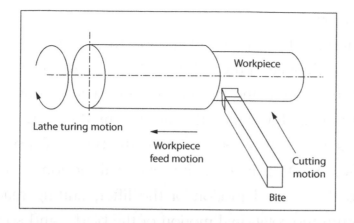

Figure 14.2   Three Kinds of Motion Made by a Lathe.

the operator simply holds on to the workpiece. Even then, the operator cannot be separated.

Here is how we could separate the operator from the lathe:

Operation 1: Return to starting position

   With conventional lathes, the operator must help guide the workpiece during processing, then must extract the processed workpiece from the lathe and set the lathe's bite and other apparatus to their starting positions to prepare the lathe for accepting another workpiece.

Operation 2: Extract processed workpiece

   The operator extracts the processed workpiece from the lathe and sets it down at the designated storage site. This is considered the next process after the lathe process.

Operation 3: Set-up the next unprocessed workpiece

   This means picking up an unprocessed workpiece and setting it up for processing. In the case of lathes, this includes setting the centering supports and the chuck supports. If the machine is a drilling machine, the operator needs to set-up the measuring jig and the V block.

Operation 4: Starting the machine

   After the operator is done setting up the lathe, he or she presses the "start" switch to begin feeding the workpiece into the lathe.

Operation 5: Processing the workpiece

In terms of the types of motion that occur, processing the workpiece in the lathe can be broken down into the cutting motion and the feed motion. The cutting motion is the speed at which the lathe turns the workpiece on the spindle. In other machines, the types of motion are different. Drills include the rotational motion of the drill and the vertical motion of the lifter; cutting machines feature the rotational motion of the blade, and so on.

Sometimes the workpiece is moved through the cutting tool, and sometimes the cutting tool is moved through the workpiece.

The above five operations can be expressed in a combination chart to help distinguish human work from machine work. (See Figure 14.3.)

As long as operations proceed as described above, there is simply no way that the operator can be completely separated from the machine. The machine must be customized to

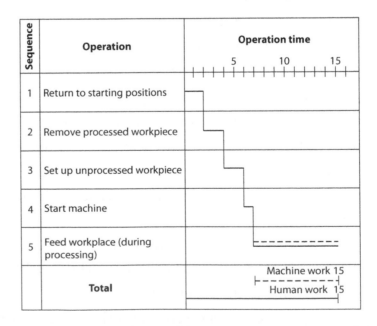

Figure 14.3 Combination Chart to Clarify Human Work from Machine Work.

enable the operator's separation. The following describes a procedure for separating the lathe operator.

Step 1: Apply *jidoka* to the cutting motion

Lathes and other cutting machines generally use rotational motion to move either the workpiece or the cutting tool. Almost all modern machines have rotational motors for automatic rotation. The rare exceptions to this are the hand-operated cutting and drilling machines that are sometimes used for woodworking.

So we generally do not have to worry about automating the rotational motion, since it is nearly always automated already. Nonetheless, we should start by considering this step and noting it on a combination chart such as the one shown in Figure 14.4.

Step 2: Apply *jidoka* to the feed motion

Once the cutting motion has been automated, we are ready to apply *jidoka* to the feed motion. For lathes, this means automating the cutting motion (as opposed to

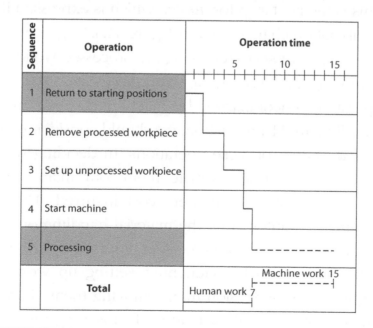

Figure 14.4 Applying Human Automation to Feeding Workpieces (to Separate the Worker).

the lathe turning motion) or workpiece feed motion. For drilling machines, it involves automating either the workpiece feed motion or the workpiece guide motion.

Once the cutting motion and the feed motion have been automated, we are able to separate the operator from the machine, at least during the processing of the workpiece. This takes us to the first stage in *jidoka*: separating the worker.

At this stage, the operator still has to extract the processed workpiece from the machine and set-up the next workpiece for processing before starting the machine. We call this pair of manual operations the "output/input" procedure or the "detach/attach" procedure. (See Figure 14.4.)

Step 3: Apply *jidoka* to the task of returning to starting position

In order for a lathe to handle processing all by itself, it must be able to fully stop both the cutting (rotational) motion and the feed motion when the processing is completed. Next, it should be able to return the cutting tool and workpiece to the positions they occupied prior to processing. This is the next step for *jidoka*, which is expressed in the combination chart shown in Figure 14.5.

Step 4: Apply *jidoka* to removing the processed workpiece

Removing and setting up workpieces are two of the operations encompassed by machine-centered material handling. In JIT production, we should consider applying *jidoka* to both of these operations. In deciding whether or not we should automate them, our main criterion is the amount of equipment cost incurred. The more complicated automating the material handling operation becomes and the more precision required of it, the more expensive it will be. Generally, setting up workpieces requires more precision than removing them. Removing them is often simply a matter of loosening the jig that holds the workpiece in place and taking the workpiece from the platform or table where it lies. Not much

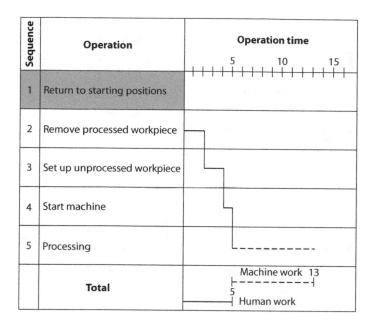

Figure 14.5 Human Automation of Return to Starting Positions (Input/Output Procedure).

precision is needed for setting down the processed workpiece either. Consequently, inexpensive devices such as pneumatic cylinders are often adequate for automating the removal of workpieces.

By contrast, it usually entails a lot more complexity and precision to set-up a workpiece into a jig or against a block correctly. Here, cheap pneumatic cylinders will not do the trick. Instead, set-up tasks usually require the precision and versatility of industrial robots. Therefore, it makes more sense to avoid trying to automate the set-up procedure if it turns out that doing it by manual labor is cheaper than buying industrial robots to do the job. Instead, we should channel our *jidoka* efforts toward the less demanding procedure of removing workpieces. (See Figure 14.6.)

Once we have automated the removal of workpieces from a machine, the operator no longer needs to remove each workpiece after setting it up and having the machine process it. This means that the operator's job

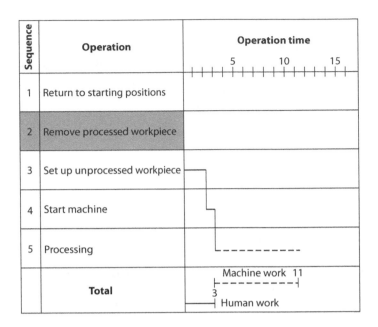

Figure 14.6 Human Automation of Removing Processed Workpieces (with Manual Set-up).

(for a series of two workpieces) changes from "remove/set-up/remove/set-up" to simply "set-up/set-up."

Step 5: Apply *jidoka* to setting up the unprocessed workpiece and starting the machine

At this point, the only remaining manual operation is setting up the workpiece and hitting the start button. Often, the same device that is able to set-up the workpiece automatically and precisely is also able to activate the machine automatically.

When a lot of precision is needed for the set-up procedure, automation may require expensive mechanisms, such as industrial robots. Therefore, we need to make a careful study of costs: Which is cheaper in the long run—manual set-up or automated set-up?

Figure 14.7 shows how the combination chart would look if we manage to automate both the set-up procedure and the machine activation procedure. As shown in the figure, this step brings the process to full automation as an "unmanned process."

Sequence	Operation	Operation time
		5　　　　10　　　　15
1	Return to starting positions	
2	Remove processed workpiece	
3	Set up unprocessed workpiece	
4	Start machine	
5	Processing	----------
	Total	Machine work 8 ----------

Figure 14.7   Human Automation of Setting Up Unprocessed Workpiece and Starting Machine (Totally Unmanned Process).

To summarize, the key points in automating processes and bringing factory automation technologies into the factory are: operators must be completely separated from the machines and the machines must be equipped with defect-detection devices, and automation must be developed one step at a time with continual attention paid to comparing manual labor costs with equipment investment costs.

It cannot be repeated enough that *jidoka* should never be used to the detriment of cost performance. Many companies have ended up taking a big loss after investing lots of money in fully automated production lines.

Case Study: Separating Workers at a Drilling Machine

In Chapter 13, we have already seen one case study of separating workers from machines. Figure 14.8 shows another example that involves a typical table-top drill wherein only the rotary motion of the drill has been automated. The operator

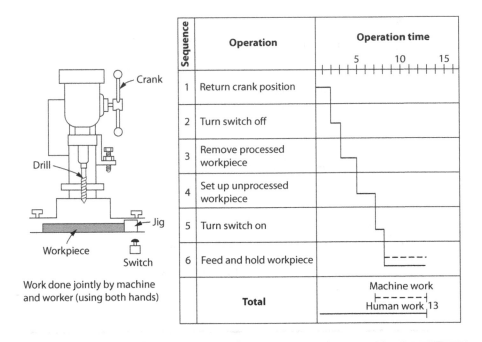

Sequence	Operation	Operation time
		5 10 15
1	Return crank position	
2	Turn switch off	
3	Remove processed workpiece	
4	Set up unprocessed workpiece	
5	Turn switch on	
6	Feed and hold workpiece	
	Total	Machine work Human work 13

Work done jointly by machine and worker (using both hands)

Figure 14.8   Table-Top Drill Operation before Improvement.

has two manual procedures to perform while using this drill: turning the crank with one hand to lower the drill to the workpiece and holding the workpiece in place with the other hand. Obviously, this drill keeps its operator busy and the operator cannot leave it at any time during the drilling process.

Improvement 1: Jidoka of "Feed"

By applying *jidoka* to the "feed" step, we can begin to separate the worker from the machine. In other words, at this stage we eliminate the need for the operator to hold the crank with his right hand and lower the drill after setting up the unprocessed workpiece and turning the start switch on. Figure 14.9 shows how the same drilling machine can be automated so that once the start switch has been pressed, the drill is automatically lowered to drill the hole, then is automatically raised back to its starting position, after which the machine shuts itself off. This frees the worker's right hand, but he still must use his left hand to hold the workpiece in place during processing. Thus, he is not completely separate from the machine.

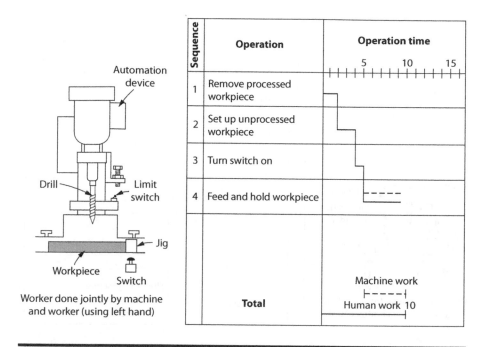

Sequence	Operation	Operation time
		5 10 15
1	Remove processed workpiece	
2	Set up unprocessed workpiece	
3	Turn switch on	
4	Feed and hold workpiece	
Total		Machine work ⊢ – – – ⊣ Human work 10

Automation device

Drill

Limit switch

Jig

Workpiece

Switch

Worker done jointly by machine and worker (using left hand)

Figure 14.9   Improvement 1: Human Automation of "Feed" Motion.

Improvement 2: Jidoka of "Hold" Motion

Our first improvement separated the worker's right hand from the machine by automating the "feed" motion. But the worker still must use his left hand to hold the workpiece in place while it is being drilled. So, he cannot be completely separated from the machine. To free both the worker's hands, we must also automate the "hold" motion that keeps his left hand busy.

Figure 14.10 shows how a pneumatic cylinder, activated by the machine's start switch, can be used to hold the workpiece in place during drilling. This enables the worker to be separate from the machine during the entire drilling operation. The worker's only remaining work is the "detach/attach" pair of tasks, in other words, removing processed workpieces and setting up unprocessed ones.

Improvement 3: Jidoka of "Detach" Movement

After the second improvement, the worker is able to be separate from the machine only while the workpiece is being

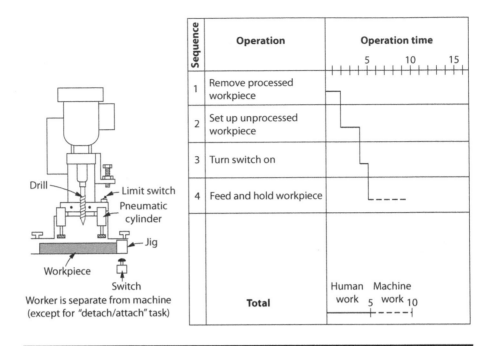

Sequence	Operation	Operation time
		5 10 15
1	Remove processed workpiece	
2	Set up unprocessed workpiece	
3	Turn switch on	
4	Feed and hold workpiece	
	Total	Human work 5 Machine work 10

Drill — Limit switch
Pneumatic cylinder
— Jig
Workpiece
Switch
Worker is separate from machine
(except for "detach/attach" task)

Figure 14.10   Improvement 2: Human Automation of "Hold" Motion.

processed (drilled). The next step is to eliminate half of the remaining pair of tasks—removing or "detaching" processed workpieces and setting up or "attaching" new ones.

Figure 14.11 shows the same drilling machine, this time with an automation device consisting of another pneumatic cylinder that pushes the processed workpiece out of the machine after the drill has returned to the starting position. The only human work remaining at this point is to set-up each workpiece in the drilling machine and press the start switch.

Ways to Prevent Defects

As mentioned earlier, it does no good to separate the worker from the machine if there is a chance that the machine will start spewing out defective goods during the worker's absence. The solution to this problem is to make the machine both capable of detecting actual or potential defects and able to shut itself off and alert operators to the problem whenever abnormalities are detected. Only then does separating

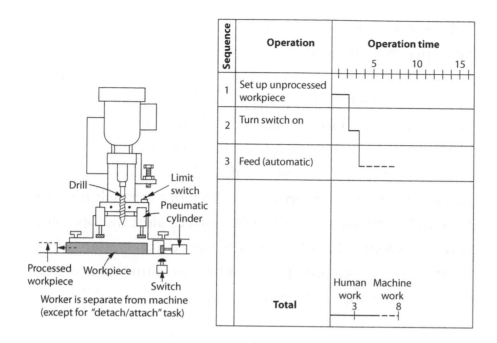

Sequence	Operation	Operation time
		5 10 15
1	Set up unprocessed workpiece	
2	Turn switch on	
3	Feed (automatic)	
Total		Human work 3 Machine work 8

Figure 14.11   Improvement 3: Human Automation of "Detach" Motion.

workers really make sense. Consequently, developing and installing defect-preventing devices is a key part of *jidoka*.

The following are a few examples of defect-preventing devices.

How to Prevent Defects in Tapping Operations

Figure 14.12 shows an example of a defect-preventing device used in tapping operations. Before this improvement, this

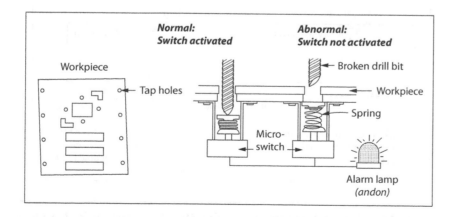

Figure 14.12   Defect-Preventing Device for Tapping Operations.

tapping machine, which uses 12 drill bits to simultaneously tap 12 places in the workpiece, experienced occasional defects such as broken drill bits, tapping omissions, and incomplete tapping. The factory had inspectors check every workpiece after being tapped to sort out the defective ones.

After the improvement, a microswitch was installed underneath each tap hole. If any of the 12 microswitches is not pressed during the tap operation, the tapping machine stops itself and lights an alarm lamp (*andon*) to alert the operators to the problem. This eliminates the need for human supervision and downstream inspection by preventing defects from recurring or being sent downstream.

How to Keep Injection Mold Burr Defects from Being Passed Downstream

Figure 14.13 shows a defect-preventing device to prevent injection mold burr defects from being passed downstream.

Before the improvement, molded workpieces were visually inspected for burr defects and were deburred when such defects were found. However, inspection oversights and other human errors occasionally resulted in the passing of workpieces with burr defects downstream. The defects went

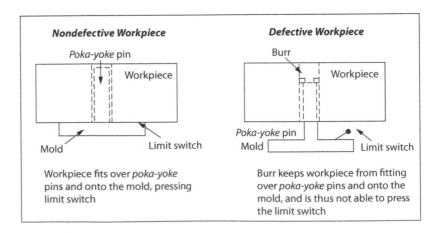

Figure 14.13   Defect-Preventing Device that Prevents Injection Mold Burr Defects from Being Passed Downstream.

unnoticed here until the final assembly stage, which caused a lot of trouble.

After the improvement, the lead wire soldering process that follows the injection molding process was equipped with a mold with *poka-yoke* pins that fit into the molded workpiece, which detected the presence of a burr in the mold and automatically stopped the lead wire soldering machine whenever one was detected. This device effectively prevents any workpieces with burr defects from reaching the final assembly process.

How to Keep Drilling Defect from Being Passed Downstream

Figure 14.14 shows a device that keeps drilling defects from being passed downstream.

This machine performs drilling and finishing in a continuous two-step process. Sometimes, however, it omits the drilling step. When this happens, the finishing drill bit breaks when trying to enter the place where the hole was omitted. Although the best thing would be to have a device that prevents drilling omissions from occurring in the first place, it was decided that it would be simpler to have a device that

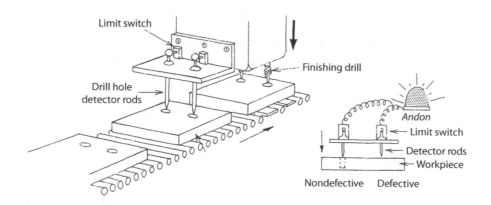

Figure 14.14 Device to Keep Drilling Defects from Being Passed Downstream.

would confirm the drilled holes just before the hole finishing step in the two-step process.

The defect-preventing device consists of a plate attached to the input side of the drill hole finishing machine. Two rods are suspended through this plate. When the drill hole finishing machine processes one workpiece, the defect-preventing device tests the next one on the conveyor by lowering the two rods through the drill holes. If a drill hole is missing, the rod cannot be lowered fully and is instead pressed back against a limit switch. When either of the limit switches are activated, the drilling and finishing machines are both stopped and an *andon* alarm is activated, as shown in Figure 14.14.

Extension of *Jidoka* to the Assembly Line

We usually apply *jidoka* to processing equipment, but we can also extend it to assembly operations to prevent defects from being passed downstream and/or to prevent overproduction. Most assembly line applications of *jidoka* are based on "A-B control" and fall into one of two categories: the full work system or the stop position system.

Full Work System

"A-B control" refers to a method for maintaining and controlling a constant flow of work by checking the passage of work between two points (A and B). The full work system helps maintain one-piece flow operations and prevents overproduction by detecting when a full workload has been reached, even when abnormalities occasionally force the line to stop. (The full work system is also discussed in Chapter 5.)

Figure 14.15 illustrates the control method used in the full work system. As can be seen in the figure, the flow of workpieces is allowed to continue only under Condition 2, in which there is a workpiece at point A but not at point B.

Point		A	B	Description
Condition	1	Yes	Yes	If there are workpieces at points A and B, moving the conveyor would cause a pile-up at point B.
	2	Yes	No	Conveyor moves only under this condition.
	3	No	Yes	If there is a workpiece at point B but not at point A, moving the conveyor would cause a gap in workpiece flow while leaving a workpiece at point B.
	4	No	No	If there are no workpieces at points A and B, moving the conveyor would simply cause a gap in workpiece flow.

Figure 14.15   A-B Control under the Full Work System.

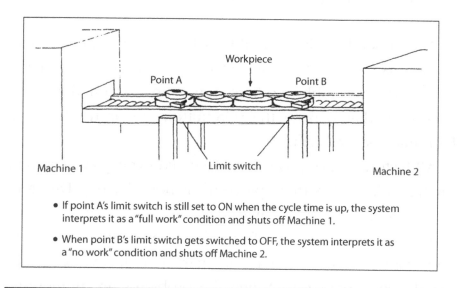

- If point A's limit switch is still set to ON when the cycle time is up, the system interprets it as a "full work" condition and shuts off Machine 1.

- When point B's limit switch gets switched to OFF, the system interprets it as a "no work" condition and shuts off Machine 2.

Figure 14.16   Full Work System Used for Machining Line.

Figure 14.16 shows an example of a full work system applied to a machining line. In this example, when the cycle time is up and the limit switch at point A is still set to ON, the system shuts off Machine 1 because producing any more goods from Machine 1 would only cause an overproduction of goods beyond the cycle time. When the limit switch at point B is switched to OFF (that is, when there are no more workpieces at point B), the system interprets this as a "no work" situation and shuts off machine B.

Stopping at Preset Positions

When an abnormality or other problem occurs on a conveyor line, such as an assembly line, the assembly workers press a stop button to stop the line immediately in order to identify the problem and solve it right way.

The following are the most common types of problems encountered on assembly lines:

1. Missing assembly part
2. Defective assembly part
3. Delay due to error in assembly method
4. Failure to keep up with assembly pitch

The assembly line should include stop buttons (also known as "SOS buttons") next to each worker. Whenever any of the assembly workers notice an abnormality, they must immediately press the SOS button to stop the line and look into the problem.

All factories have problems. We could even go as far as to say that a factory without problems is not a factory. Different problems crop up from day to day.

The same goes for the factory's assembly line. Assembly line problems range from missing parts to defective parts and unbalanced operations. When the problems are numerous, pressing the SOS button each time may result in a line that is almost always stopped, which is counterproductive.

Although it is important to stop the line to identify and solve the problems, line supervisors believe it is equally, if not more, important for the line to operate smoothly and productively.

The system of stopping at preset positions is a good way to find a middle path through the mixed intentions of supervisors who want the line stopped in order to identify and solve problems, but who also want to keep the line running productively.

Figure 14.17 shows this system being used for an assembly conveyor line.

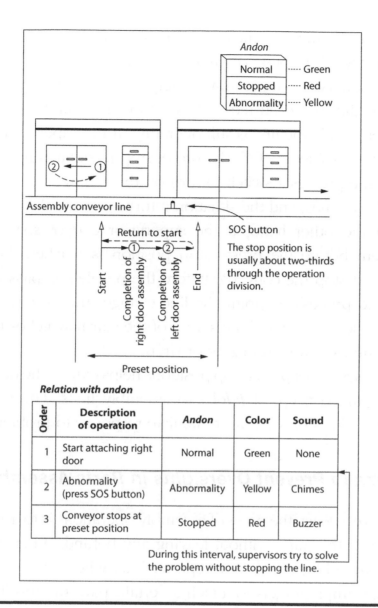

Figure 14.17 Stopping at Preset Positions on an Assembly Conveyor Line.

Let us suppose an assembly worker has just started an assembly operation and is about to fasten the right door onto the product. While doing this, the worker notices an abnormality and immediately presses the nearby SOS button, which is usually located about two-thirds the way along the path covered by the assembly worker during the assembly operation.

Once this worker presses the SOS button, the *andon* changes color from green (normal) to yellow (abnormality). Usually a number identifying the specific process along the assembly line is displayed, and a chime or bell rings to alert the supervisor. (For further description of *andon,* see Chapter 9.)

The supervisor comes immediately to the process where the abnormality has occurred and tries to identify and solve the problem while the line is still operating. If the supervisor can solve the problem before the preset stop position is reached, he or she presses a switch to turn off the yellow *andon* light and the chimes, and the situation returns to normal.

On the other hand, if the supervisor cannot solve the problem before the preset stop position is reached, he or she must stop the conveyor before the problem is passed to the next process. Stopping the line changes the *andon* color from yellow to red and the sound of the alarm switches from soft chimes to a loud buzzer or siren.

This system of preset stop positions helps extend the defect-preventing concept of *jidoka* to assembly lines. The preset stop positions provide an immediate response to problems.

Jidoka *to Prevent Oversights in Parts Assembly*

At the very least, the point of assembly operations is to assemble all of the parts without leaving any behind. When even this basic obligation is not kept, such as when an assembly worker simply forgets to attach a certain part, the result is a defective product. This is where *poka-yoke* devices can be used as an extension of *jidoka* to prevent such defects that arise from the omission of parts. (For further descriptions of *poka-yoke* devices, see Chapter 12 of this manual.)

Figure 14.18 shows an example of this extension of *jidoka* to prevent the omission of a parts tightening operation. Before the improvement, the assembly worker used an impact wrench to tighten the fasteners in the workpieces being assembled. Occasionally, the worker would forget to perform this fastening operation, and naturally the result was a defective product.

Instead of relying on the worker's memory and vision to use the impact wrench to tighten the workpieces, a pneumatic

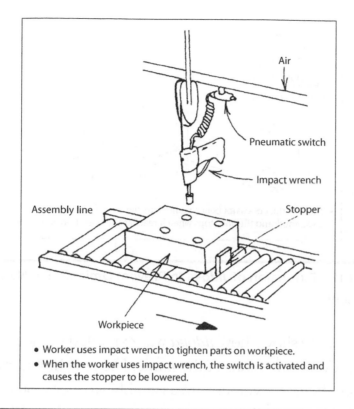

Air

Pneumatic switch

Impact wrench

Assembly line

Stopper

Workpiece

- Worker uses impact wrench to tighten parts on workpiece.
- When the worker uses impact wrench, the switch is activated and causes the stopper to be lowered.

Figure 14.18   Extension of *Jidoka* to Prevent Omission of Workpiece Parts Tightening.

switch was installed. When the worker uses the impact wrench, the switch is activated, which causes the stopper to be lowered so the workpiece can continue on the conveyor. If the worker forgets to use the impact wrench, the stopper holds the workpiece in place. This device reduced the number of untightened workpieces to zero.

Another *Jidoka to Prevent Oversights in Attaching Nameplates*

One of the basic requirements for productive assembly line operations is to keep operations level, well-ordered, and within the cycle time. If the operational procedures are allowed to vary between one workpiece and the next, or if the workers are allowed to use their own discretion concerning how to do things, the assembly line is bound to produce products with missing or improperly assembled parts.

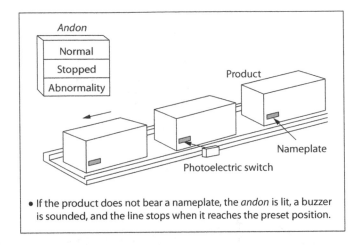

• If the product does not bear a nameplate, the *andon* is lit, a buzzer is sounded, and the line stops when it reaches the preset position.

Figure 14.19   Extension of *Jidoka* to Prevent Omission of Nameplate Attachment.

Figure 14.19 shows how *jidoka* was extended to the assembly line to prevent omissions at the nameplate attachment process.

Before the improvement, an assembly worker would sometimes overlook attaching a nameplate to a product. This happened more often when the worker had just come back from a break. When this problem was first noticed, the supervisor made it a point to remind workers to be careful about attaching nameplates to every product. Still, workers occasionally forgot. Finally, the supervisor decided the assembly line should have a *poka-yoke* device that would prevent products without nameplates from proceeding down the line.

The *poka-yoke* device consists of a photoelectric switch that reflects a light beam off of the shiny metal nameplate. This switch uses the reflected beam to detect whether the nameplate has been attached. If it detects a missing nameplate, it lights the "abnormality" *andon* and sounds a buzzer. The line is not stopped until the product reaches a preset position. This device prevented any more products from being shipped without nameplates.

Chapter 15

Maintenance and Safety

Existing Maintenance Conditions on the Factory Floor

I have met many factory managers who pretty much accept machine breakdowns as part of the inevitable facts of factory life. But when I look around at their factories, I invariably notice at least some of the following conditions:

- Floors dirtied by puddles of oil leaked from machines
- Metal shavings scattered all over machines and the floor
- Machines so dirty that people avoid touching them
- Clogged air ducts that emit dust into the room
- Level gauges so dirty that they are hard to read
- Oil and dirt around the oil inlet ports
- Muddy oil in the oil tanks
- Leaks in the hydraulic and pneumatic equipment
- Loose bolts and nuts
- Strange noises coming from machines
- Machines vibrating abnormally
- Dirt and dust piled up on the photoelectric sensors and limit switches
- Abnormally hot motors
- Sparks flying from shorted wires
- Loose V belts

- Damaged V belts still being used
- Broken gauges and measuring instruments still being used
- Cracks filled with cardboard, jerry-rigging, and other temporary repairs

It was not at all hard to come up with this list of nearly 20 objectionable conditions. In fact, this list is based only on my observations in and around factory equipment; it would be a much longer list if I included all the other undesirable conditions I have run into in other parts of factories.

When I look around a factory and see many of these conditions existing, I can tell that JIT production was never even attempted there. Whether the factory uses small machines or large ones, there is no excuse for breakdowns. As I have mentioned elsewhere in this manual, factory managers need to emphasize the equipment's possible utilization rate over its capacity utilization rate.

The following pages explain why JIT production insists on zero breakdowns.

What Is Maintenance?

Why Is "Possible Utilization Rate" Necessary?

One way to look at JIT production is to compare it to the body's circulatory system, in which the blood flows to the various organs "just-in-time" to be used. Just as the factory handles large and small parts for its products, so too does the body have its large arteries and small veins and capillaries.

In JIT production, however, any delay in the flow of small parts (in the "veins" or processing line) soon stops the flow of large parts (in the "arteries" or assembly line).

To prevent such problems, JIT production vitally depends on maintaining a condition of zero breakdowns. This makes proper maintenance an essential part of JIT production. That is why it is more important to maximize the equipment's "possible utilization rate" (the availability of functioning equipment) than to raise its capacity utilization rate. People need to know the equipment will be in working order whenever they need it.

The key to achieving zero breakdowns is not maintenance in terms of repairing broken down equipment, but rather "preventive maintenance" that treats the causes of breakdowns before the breakdowns actually happen.

Why Accidents Happen

Why do accidents happen? The simplest and most direct answer is "deterioration." From the day a machine is installed, its condition gradually deteriorates over years of use, and sooner or later the combination of deteriorated parts or the accumulated deterioration of a single part will cause the machine to break down.

Almost any machine will have some telltale symptoms of ill health before it actually breaks down. For example, the machine may no longer be able to meet the required quality standards and may stop intermittently. Figure 15.1 shows the downhill path most machines follow before breaking down.

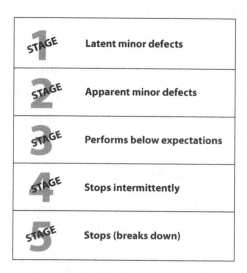

Figure 15.1   Stages on the Path to Equipment Breakdown.

The important thing is to learn to recognize where each machine is on that path.

Stage 1: Latent Minor Defects

Though difficult to see or hear, the machine's rotating parts are operating under increasing friction and its fastened parts are getting a little looser. These and other subtle defects characterize the first stage of equipment deterioration.

Stage 2: Apparent Minor Defects

The same defects described in the first stage have now become somewhat noticeable to the eye or ear. In addition, the machine may be vibrating more, making more noise, and leaking small amounts of oil, water, or air. But none of these defects are major enough to impair the machine's functioning.

Stage 3: Performs below Expectations

At this stage, it has become difficult to get the machine to perform with the desired precision and within the dimensional tolerances. The machine is turning out products with widely varying quality and suddenly it needs more adjusting than it used to require. It can no longer keep up with quality standards and is producing lower yields.

Stage 4: Stops Intermittently

At this stage, the machine has to be shut off fairly often to make adjustments to bring the product quality back into line. The machine frequently turns out damaged or dented goods, but can usually be started up again after making simple adjustments or repairs.

Stage 5: Stops or Breaks Down

At this final stage, the machine functions so poorly that it stops itself, which is to say it breaks down.

We should keep in mind that machines usually break down due to deterioration, and these kinds of breakdowns never happen all of a sudden; they happen in stages. One or more of the machine's deteriorating parts are left to deteriorate and eventually this deterioration accumulates or combines in a simple or complicated way to cause a breakdown.

If we respond to deterioration only when it reaches the fifth stage, we still will have to deal soon with various machines that are currently at the other four stages in the path. In other words, we cannot hope for a true reduction in breakdowns until we work our way up the path and treat deterioration before it results in breakdowns.

Maintenance Campaigns

When we let factory equipment deteriorate, sooner or later it will break down. In view of this, how can we achieve zero breakdowns? We must take measures to slow or halt equipment deterioration before it reaches the breakdown stage.

In JIT production, we do this by promoting and establishing a cycle of four basic maintenance activities within the staff hierarchy of each company division. Figure 15.2 illustrates this fourfold company-wide approach.

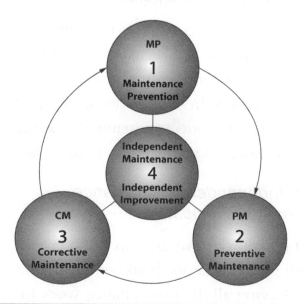

Figure 15.2   Production Maintenance Cycle for Zero Breakdowns and Zero Defects.

Measure 1: Maintenance Prevention (MP)

Maintenance prevention mainly pertains to equipment design. It involves using the data provided by independent maintenance and independent improvement activities to design equipment that is less likely to break down or experience faulty operation, and is more conducive to deterioration-preventive measures. Another important design criterion that is influenced by MP is the challenge to make equipment that can be maintained more easily, more quickly, correctly, and safely.

Measure 2: Preventive Maintenance (PM)

Preventive maintenance centers on daily checking and maintenance procedures that form part of independent maintenance and independent improvement activities. It also seeks to raise the reliability of the equipment while reducing the risk of faulty operation and slowing the progress of equipment deterioration. In addition, PM involves studying and selecting operational methods and equipment to help make maintenance activities easier to perform.

Measure 3: Corrective Maintenance (CM)

Corrective maintenance comprises the maintenance procedures taken in response to a breakdown, with a view toward preventing the problem's recurrence and improving the equipment's condition. In addition to reversing deterioration and raising reliability, corrective maintenance seeks to make the equipment easier to maintain on a daily basis.

Measure 4: Independent Maintenance, Independent Improvement

To reduce breakdowns, we give up the conventional notion that the equipment operators should simply operate the equipment while leaving all the maintenance work to the maintenance technicians. After all, the equipment operators are the ones who know the equipment best—they are the first to notice when the machine's motor starts sounding funny or when formerly clean parts of the machine are streaked with oil or dirt. Equipment operators should embrace with pride the idea that they can take care of their own machines. They should put that concept into practice by cleaning, checking, and oiling their machines. They can even replace parts and perform minor repairs.

Meanwhile, the maintenance technicians can still play an important role by promoting and teaching accurate and prompt repair methods to the equipment operators for improved independent maintenance and independent improvement activities. In so doing, they can help make the whole MP-PM-CM cycle run more smoothly.

CCO: Three Lessons in Maintenance

These days, when JIT consultants describe how to maintain a neat and orderly factory, they find it difficult to limit the basics to just five (the 5S's). Some list 6S's and others 7S's. Adding more S's is not always an improvement. Nonetheless, many Japanese companies are inclined to include *shukan* (custom) as the sixth S.

For our purposes, let us recognize that implementing and enforcing the 5S's daily is a good practice for companies. This is especially true when it comes to the 5S's as they relate to equipment maintenance.

In particular, equipment maintenance activities should include three main customs: Cleanliness, Checking, and Oiling (CCO). We refer to them together this way because they should always be carried out as a threefold unit that forms the core of independent maintenance activities.

Let us take a closer look at each part of the CCO formula.

Cleanliness (C)

As part of the 5S's, cleanliness (*seiso*) is the routine housekeeping work that is essential for maintaining the day-to-day health of the factory. As applied specifically to equipment, maintaining cleanliness is the best way to make a daily examination of the equipment. (Cleanliness is described in more detail in Chapter 4.)

Unfortunately, once people have cleaned up their workshop, they let it go for days, offering such excuses as, "We're too busy to get to that right now" or "Hey, it's still clean."

Sometimes it is the workshop supervisor who causes problems. For example, a supervisor might insist that cleanliness tasks be performed outside of regular working hours or that daily cleanliness activities do not improve productivity enough to be worth the trouble. But the fact remains that cleanliness will never lead to zero defects and zero breakdowns unless it is kept up as an integral part of daily production activities.

First of all, maintaining cleanliness is not something to be done at the odd moment between one production operation and the next. Instead, we should view it as an essential part of preproduction activities, just like changeover prior to processing a new model or setting up parts trays before assembling a new model.

In other words, equipment operators need to fully recognize the importance of maintaining cleanliness and make it (along with checking and oiling) just as much a part of their daily routine as anything else they do day in and day out in the factory.

To help operators stay on top of their CCO duties, workshop supervisors should post a "cleanliness inspection checklist" in the workshop, which operators can use to keep track of how well the daily cleaning tasks are being carried out. (This checklist is shown in Chapters 4 and 16.)

Just as each workshop should have tools and other equipment reserved expressly for changeover operations, so should it include the specific tools necessary to maintain cleanliness.

Checking (C)

Maintenance should be understood as an activity designed to prevent equipment from breaking down. The purpose of checking, therefore, is to determine whether the equipment is about to break down.

Checking is undeniably part of maintenance activities—but not something to be left entirely up to the maintenance technician. Since the operator is the one who knows best how well or poorly the equipment is operating, the operator has the kind of concrete problem-consciousness needed for effective daily checking and, when necessary, prompt response.

In recognition of the operator's superior qualifications as an equipment checker, we should not downplay his or her checking duties by relegating them to "spare time" or "overtime." They must be clearly established as part and parcel of the operator's daily routine.

Figure 15.3 shows a cleanliness inspection checklist and some cleanliness check cards. In this example, the workshop also includes a "cleanliness control board" on which operators post cleanliness check cards. The cards note whether the check ended normally or whether an abnormality was found. This control board enables the supervisor to immediately understand whenever an abnormality is found, so that a prompt response can be made.

Oiling (O)

The Just-In-Time concept of "just what is needed, just when it is needed, and just in the amount needed" can be applied directly to the activity of oiling. In other words, we need to give each machine just the kind of oil it needs, just when it

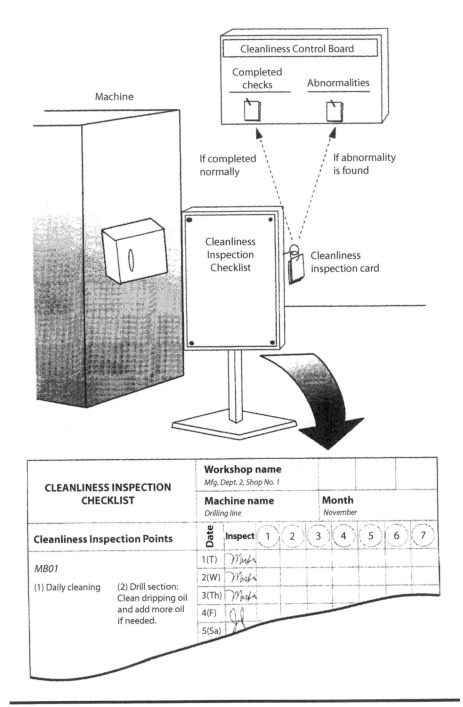

Figure 15.3 Cleanliness Inspection Checklist and Cleanliness Check Cards.

is needed, and in just the amount needed. (Proper oiling is also discussed in Chapter 4.)

The management of this activity should be made as visible as possible so that everyone can understand it. Figure 15.4 shows how the visual control tool known as *kanban* can be used

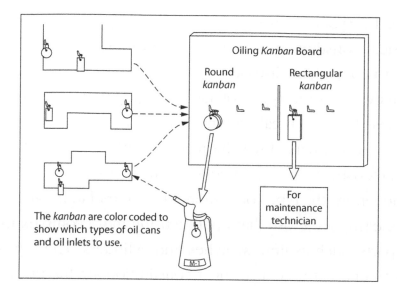

Figure 15.4 *Kanban* **for Oiling.**

to indicate what kind of oil goes where. These *kanban* also employ another visual control method known as color coding.

Here is how the *kanban* are used in the example shown in Figure 15.4.

1. Separate *kanban* are established for each machine and each oil inlet port.
2. Round *kanban* indicate oiling done by the workshop supervisor and rectangular ones indicate oiling done by the maintenance technician.
3. The *kanban* are color coded to indicate which type of oil and which inlet port to use, and to mark other material, such as oil cans and oiling tools.
4. The oiling times and amounts used are entered on the inspection checklist or in a log book.

Preventing Breakdowns

Some people are stronger than others. Some people catch colds easily while others can go all year without even a stuffy nose. Everyone knows that different people have different physical constitutions that make them more or less susceptible to contagious diseases.

Likewise, some types of factory equipment are stronger and less likely to break down, while other types are weaker and tend to break down more easily. We can refer to this characteristic as the equipment's "constitution."

Generally, the types of equipment that tend to break down more easily are those that operate using more complex moving parts, such as limit switches and cylinders. The types of equipment that have a stronger constitution are the ones that operate using simple coupling devices, such as cams and gears. It is also much less obvious when limit switches and cylinders are not operating correctly than when gears go on the blink.

Figure 15.5 shows two devices for holding down workpieces in a drilling machine. One device is a pair of pneumatic cylinders. If either of the pneumatic cylinders malfunctions, there is a safety hazard in that the cylinder might begin to operate while the worker is still setting up the workpiece, and the worker could get a pinched hand. For safety reasons, it makes more sense to use the other device, which is simply a pair of springs.

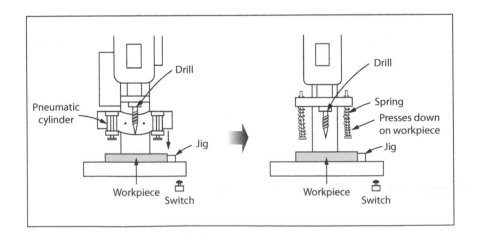

Figure 15.5 Safety Improvement from Pneumatic Cylinders to Springs.

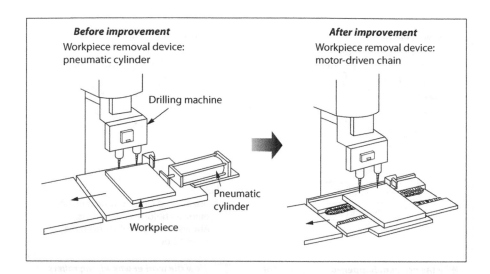

Figure 15.6 Use of Motor-Driven Chain as Automatic Workpiece Removal Device.

Figure 15.6 shows an improvement made in the method of automatically removing processed workpieces from a drilling machine. To facilitate maintenance and reduce defects, the workpiece removing device was changed from a cylinder to a motor-driven chain.

Once a breakdown occurs, we must find the cause and make an improvement that will prevent the same kind of breakdown from occurring again. To do this, the people who are dealing with the breakdown must see it first-hand, get the data first-hand, and then make a decision about how to respond effectively to the problem. Stopgap measures are not the answer. Whatever is done to fix the problem must be a preventive measure, not just a temporary patch job.

Why Do Injuries Occur?

It is pretty safe to say that every factory has at least one "Safety First" type of sign or banner on display. Factor managers and employees are conscious of the need for assuring safety, but accidents still happen, and they often happen

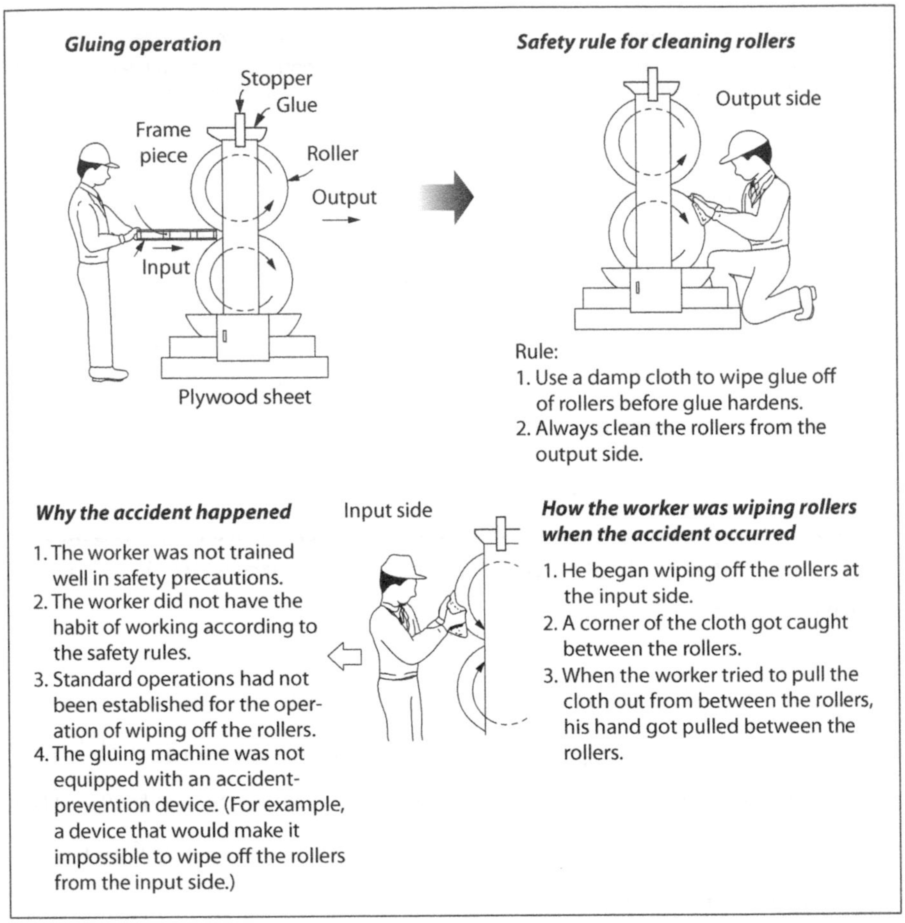

Figure 15.7   An Accident at a Plywood Gluing Process.

when a machine breaks down. If people want to give more than lip service to safety, they must address the need to prevent breakdowns.

Figure 15.7 shows an example of how an accident occurred during a plywood gluing operation. Naturally, the factory where this happened was not without its "Safety First" banner.

The accident actually happened at the end of the day, when a worker was cleaning the glue roller that presses together the sheets of plywood. As soon as the last set of plywood sheets was pressed, the worker took a damp cloth and began holding it against the rotating rollers to wipe off the excess glue before it hardened. The worker did this from the same side he had input the plywood sheets—a violation of the safety rule stating that the rollers must always be cleaned from the output side.

The worker broke this rule as a matter of habit. As shown in Figure 15.7, the rollers rotate in opposite directions to press the plywood between them. When wiping the rollers at the input side, an edge of the cloth would sometimes get pulled between the rollers. The worker relied on his reflexes to pull the cloth back before the rollers got a good grip on it. In other words, the worker gave higher priority to his reflexes than to the concept of "safety first." In hindsight, it seemed obvious to everyone that the worker's behavior would eventually lead to an accident.

The only way to effectively prevent this kind of accident from happening again is to clarify just why it occurred and take every countermeasure necessary to prevent a recurrence.

The main reason for this accident's occurrence include the following:

1. The worker was not adequately trained to be aware of the dangers inherent in his job and to take safety precautions.
2. The safety rule saying that workers must wipe the rollers from the output side was put into the book, but not into the mind of the worker. The supervisor is responsible for seeing to it that workers make a habit of obeying the rules.
3. Safety had not been built into the operational procedures. The way to do this is by establishing safety-conscious standard operations.
4. The equipment lacked an accident-prevention device, such as boards installed just in front of the rollers on the input side that would block access to the rollers for wiping. The worker would then be required to wipe the rollers from the output side.

What Is Safety?

Factory managers are faced with many ongoing needs, such as the need to raise productivity and improve quality. However, no need should ever take priority over the need to assure safety.

In other words, no boost in productivity or quality can ever be justified if it is at the expense of safety. Safety is everything in manufacturing—it is where manufacturing must start and end.

You would not know this judging from the kinds of excuses workers give after an accident and/or injury. Some say, "I was daydreaming" or "I was hurrying to catch up." Workshop supervisors must speak the plain truth and make it known when the rules are bent or broken, or when workers fail to make a habit of doing things the safe way.

Another way to prevent accidents is to develop devices that make it difficult, if not impossible, to "daydream" or "hurry up" at safety's expense. Rather than simply dispensing tongue lashings after accidents occur, supervisors should take preventive action by checking up regularly on safety practices and sternly warning workers who fail to obey the safety rules. After all, the correct or incorrect behavior of factory workers is a direct reflection upon the ability of the supervisors and factory managers to carry out their duties responsibly. Achieving zero injuries and zero accidents is a goal the entire company should pursue together, and a key part of such a company-wide safety campaign is devising ways to prevent shop-floor injuries and accidents.

Let us review the accident example shown in Figure 15.7 and the lessons to be learned from that incident. The following summarizes the four improvement points to be made to prevent similar accidents from recurring.

1. Establish more complete basic training
   The entire training program needs to be reviewed and improved so that workers are taught not only about the flow of goods in the factory and the features of

the equipment, but also about the proper attitude and approach toward safety assurance.

2. Get into the habit of obeying the rules

   Workers should make maintaining the 5S's and following the safety rules so habitual that they rarely need to think about it. When safety assurance requires that workers use their hands and voices to keep each other informed of what is happening, such behavior must become a natural habit. Workshop supervisors need to be especially strict in enforcing this.

3. Establish standard operations

   Along with training to teach the habit of obeying the rules, establishing safety-conscious standard operations and maintaining them with visual control tools will enable anyone to understand how things should be done. It will help supervisors keep tabs on whether operations are being done by the book.

4. Develop devices that prevent injuries and accidents

   No matter how well the rules are taught and enforced, people will occasionally make mistakes. We can still help prevent injuries and accidents that arise from human error by developing devices that make it difficult or impossible to err in an unsafe manner. We have seen how *poka-yoke* devices can prevent defects from being produced. We must extend the *poka-yoke* concept and create "safety *poka-yoke*" devices that prevent accidents.

Strategies for Zero Injuries and Zero Accidents

Thorough Implementation of Standard Operations and Rules

The first principle in safety assurance is to establish and maintain standards. The lion's share of injuries and other accidents occur when something is done in a nonstandard and abnormal manner.

We use standards to clearly distinguish between what is normal and what is abnormal. In factories, we should use visual control methods to make it obvious to anyone when things are nonstandard and abnormal.

Orderliness (*seiton*) calls for the creation of standard locations for items to assure safety in the physical layout of the factory. Likewise, standard operations require the creation of operation standards to help eliminate injuries and accidents. Standard operations are like the pillar supporting safe operations and training workers to maintain standard operations is like a crossbeam connected to that pillar. Together, they provide the main support for the structure of production operations. The point of this analogy is to underscore the importance of standards for factory layout and production operations.

Figure 15.8 shows a standard operations chart marked with crosses at all key safety points. Of course, the specific safety standards are described in the standard operations manual and operations guide to keep workers informed of safety-conscious procedures and safety precautions. Each company needs to invest enough resources to thoroughly educate and train workers in standard operations that help assure safety.

The more workers must assist in machine work, the greater the risk of injury. Therefore, the separation of workers from machines achieved through *jidoka* can be an important contribution to safety. (For a further description of how *jidoka* separates workers, see Chapter 14.)

Obviously, separating workers from machines that use sharp tools, such as saw blades or drill bits, helps to assure safety. The same goes for presses and other manufacturing equipment. Figure 15.9 shows how the worker was separated from the machine in the case of a lathe used for punching

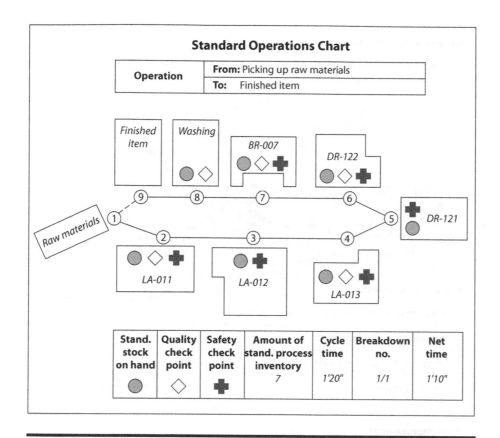

Figure 15.8 Standard Operations Chart Marked with Safety Points.

holes. Before the improvement, the lathe operator had to control the cutting motion and set the lathe back to the starting position. This kept him at the machine and kept productivity at a rather low level. Moreover, it exposed the operator to risk of injury from the rotating hole-punching bar and other moving parts of the lathe.

After the improvement, a hydraulic cylinder was used to control the cutting motion and the position setback was also automated, thereby enabling the operator's separation from the lathe. This not only significantly boosted safety assurance, but also doubled productivity

Another safety-enhancing improvement having to do with presses is the simple relocation of start buttons. Figure 15.10 shows a group of five presses handled by a single worker in a U-shaped manufacturing cell using multi-process operations.

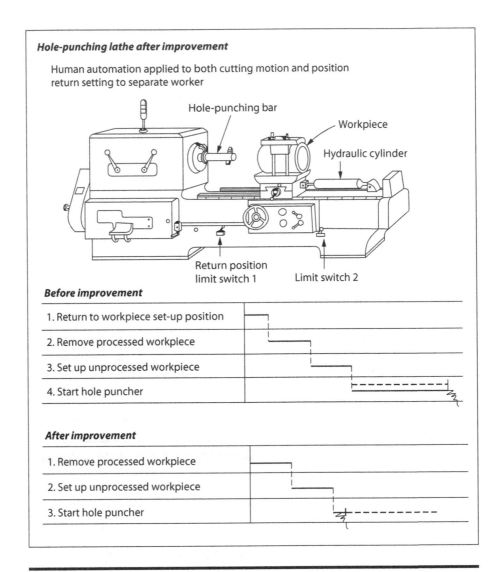

Hole-punching lathe after improvement

Human automation applied to both cutting motion and position return setting to separate worker

Hole-punching bar

Workpiece

Hydraulic cylinder

Return position limit switch 1

Limit switch 2

Before improvement

1. Return to workpiece set-up position
2. Remove processed workpiece
3. Set up unprocessed workpiece
4. Start hole puncher

After improvement

1. Remove processed workpiece
2. Set up unprocessed workpiece
3. Start hole puncher

Figure 15.9   Separation of a Worker from a Hole-Punching Lathe.

The start button on each press was moved to the next press in the cell so that the worker can start the previous press as he comes to the next one and is always at a safe distance from the press when it starts operating.

A common safety problem with presses is that sometimes, just after the operator sets up the workpiece and presses the start button, he notices the workpiece is slightly out of position and, without thinking, tries to quickly correct it before the press comes down—a sort of "reflex" response that often leads to accidents. Obviously, nobody gets injured intentionally, but sometimes workers let their reflexes overcome their

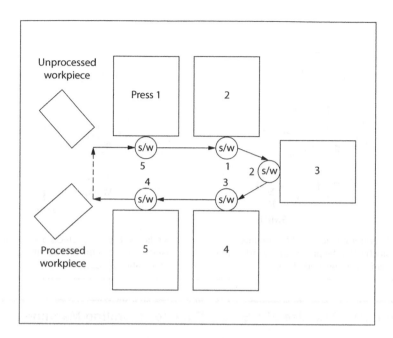

Figure 15.10 Separation of a Worker from Presses.

rational judgment. This is another good reason for separating workers from machines whenever possible.

Poka-Yoke *Applied to Safety*

Poka-yoke devices are mistake-proofing devices that can work to prevent defects or, in this case, accidents and injuries. Since careless human behavior is a leading cause of accidents, safety *poka-yoke* devices can provide a very effective means of preventing accidents.

The following are a few examples of safety *poka-yoke* devices.

Attaching a Safety Plate to a Drilling Machine

Generally, workers are not allowed to wear work gloves when operating drilling machines because it increases the danger of injury from the spinning drill. Figure 15.11 shows how attaching an acrylic safety plate in front of the drill not only enables the operator to avoid touching the drill bit, but also prevents him from getting his hands pinched by the pneumatic cylinders holding down the workpiece.

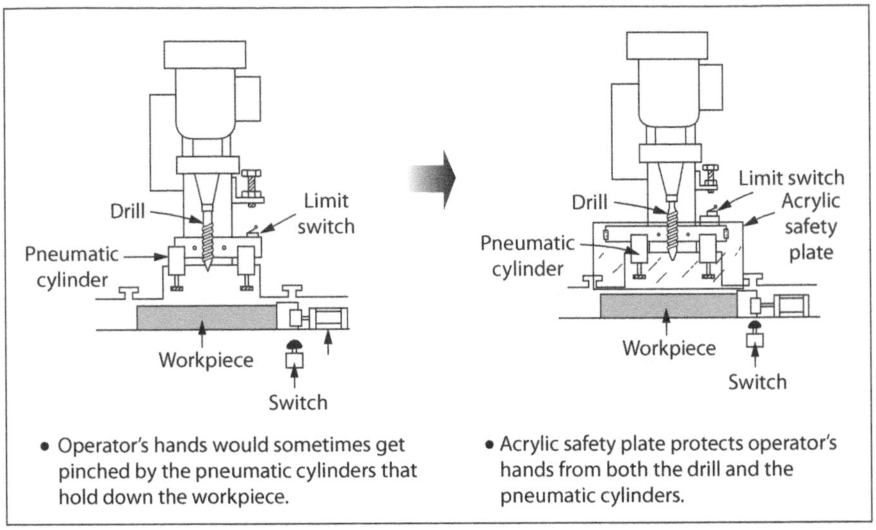
Drill
Limit switch
Pneumatic cylinder
Workpiece
Switch
• Operator's hands would sometimes get pinched by the pneumatic cylinders that hold down the workpiece.
Drill
Limit switch
Acrylic safety plate
Pneumatic cylinder
Workpiece
Switch
• Acrylic safety plate protects operator's hands from both the drill and the pneumatic cylinders.

Figure 15.11 The Use of a Safety Plate for a Drilling Machine.

Safety Cage on a Press

Presses cause more injuries than most other types of manufacturing equipment. As described earlier, presses tempt their operators to act on reflex rather than on reason. As a result, many presses are equipped with start switches that require two-hand operation. Some also have "electronic eyes" that shut off the press if any foreign object intrudes into the danger zone.

The best safety device is one that enables complete separation of the worker from the press, since it allows the worker to remove himself from the press area while the press is operating. While safety is more important than productivity, it is obviously much better to find a way of ensuring total safety without sacrificing productivity. The best devices improve both safety and productivity.

Figure 15.12 shows a press upon which a safety cage was installed. The operator sets up the workpiece, shuts the cage door, and then starts the press. Once he shuts the cage door, the operator is completely cut off from the press. Using this safety cage is better than using a two-hand start switch since it enables the operator to be separated from the press, which boosts productivity by freeing the operator for other tasks.

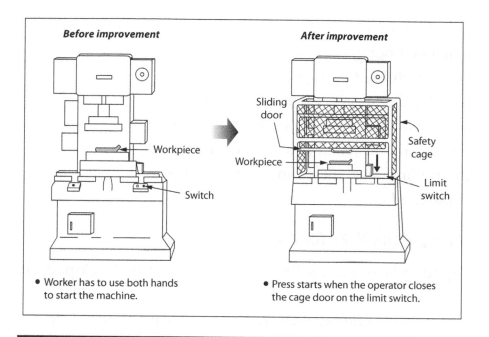

Before improvement

Workpiece

Switch

- Worker has to use both hands to start the machine.

After improvement

Sliding door

Workpiece

Safety cage

Limit switch

- Press starts when the operator closes the cage door on the limit switch.

Figure 15.12   A Press with a Safety Cage.

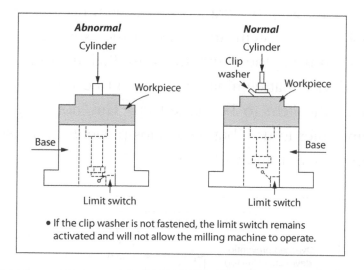

Abnormal

Cylinder

Workpiece

Base

Limit switch

Normal

Cylinder

Clip washer

Workpiece

Base

Limit switch

- If the clip washer is not fastened, the limit switch remains activated and will not allow the milling machine to operate.

Figure 15.13   A Safety *Poka-Yoke* Device for a Milling Machine.

Safety Poka-Yoke *for a Milling Machine*

Figure 15.13 shows a safety *poka-yoke* that was developed for a milling machine.

When operating the milling machine, the operator first sets up the workpiece, then uses a clip washer to hold the workpiece in place before starting the milling machine. If the operator ever forgets the washer, the workpiece can be ejected from the machine, which is dangerous indeed. Milling machine operators were warned of this danger and told to be very careful not to forget.

After the improvement, a limit switch was installed in the base and the cylinder presses upon this switch unless it is held by the fastened clip washer. This limit switch prevents the milling machine from operating unless the clip washer is fastened.

Safety Poka-Yoke *for a Crane*

Figure 15.14 shows a safety *poka-yoke* that was developed for a crane.

The crane's rail was not well reinforced and therefore had a rather modest load capacity. Overloading the crane was very dangerous, but workers seldom took the trouble to weigh things before using the crane to pick them up. Instead, they just looked at the item and guessed the weight. To assure safety when using the crane, they installed an overload prevention device that eliminated the need to even estimate the weight of the item to be picked up. This not only makes the crane safer to use, but also helps prevent the hoist from breaking down from overloading.

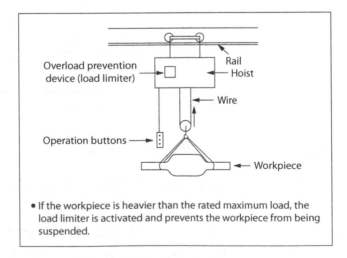

- If the workpiece is heavier than the rated maximum load, the load limiter is activated and prevents the workpiece from being suspended.

Figure 15.14 A Safety *Poka-Yoke* Device for a Crane.

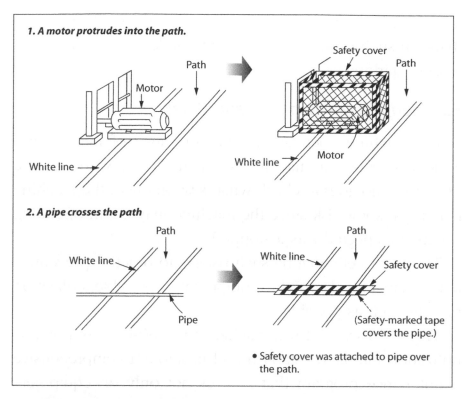

1. A motor protrudes into the path.

Path

Motor

White line

Safety cover

Path

Motor

White line

2. A pipe crosses the path

Path

White line

Pipe

Path

White line

Safety cover

(Safety-marked tape covers the pipe.)

● Safety cover was attached to pipe over the path.

3. A servo shaft protrudes into the path.

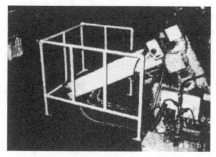

● A safety rail was installed around the servo shaft.

Figure 15.15   Visual Safety Assurance.

Visual Safety Assurance

Figure 15.15 shows an example of visual safety assurance. In this factory, the path that workers use to get around the factory contains obstacles, such as transversing pipes and protruding pieces of equipment. The best improvement would be to find some way to reroute the pipes and move the equipment out of the path. But for practical reasons the factory decided on a second-best improvement, which was to mark all such obstacles with easily visible safety covers and hazard markings.

Full-Fledged Maintenance and Safety

Faulty machine operation is another cause of injuries and accidents. When a machine suddenly stops working, the operator who goes to check what is wrong with the machine may be at some risk since the machine may begin operating just as unexpectedly as it stopped.

The key to eliminating such risks is to practice preventive maintenance to keep the equipment's "possible utilization rate" as high as possible.

The way to work toward achieving a 100-percent possible utilization rate is to establish and promote a comprehensive maintenance program that focuses not only on equipment operators, but the entire company. It must include the following two features:

1. *Thorough training in CCO.*
   Cleanliness, Checking, and Oiling (CCO) must become a daily habit for all equipment operators—an integral part of their routine tasks.
2. *Development of machines with "strong constitutions" that do not easily break down.*
   Some types of machines are weaker in "constitution" than others. As mentioned earlier, machines that operate using limit switches and cylinders are weaker than those that operate using direct coupling devices, such as gears and cams. Whenever possible, if we use the stronger types of drive mechanisms to do the job, we will find our machines less likely to break down.

In summary, various ways of improving safety assurance have been discussed including: preventive maintenance in pursuit of a 100-percent possible utilization rate, standardization of operations, full-fledged safety training, wider use of visual and audio safety-enhancing devices throughout the factory, and safety-oriented *poka-yoke* devices.

When all is said and done, safety is our main concern.

Index

1973 oil crisis, 8
3 Mu's, 643
    eliminating, 151
5 Whys and 1 How, 24, 128, 129, 130–134
    waste discovery by, 208–210
5MQS waste, 152–153
    conveyor waste, 155–156
    disaster prevention measures waste, 159
    large machines waste, 154–155
    materials waste, 157
    parts waste, 157
    searching waste, 154
    shish-kabob production waste, 158
    walking waste, 153–154
    waste in air-processing machines, 156–157
    waste in defective goods production, 159
    waste in meetings, 158
    watching waste, 154
    workpiece motion waste, 158–159
5S approach, xii, 230, 237–238, 455, 689, 721
    as bridge to other improvements, 264
    as prerequisite for flow production, 344
    benefits, 238–243
    changeover 5S checklist, 512
    for factory improvement, 15–17
    in changeover procedure improvement, 502
    keys to success, 262–264
    meaning, 243–249, 250
    orderliness applied to jigs and tools, 307–319
    red tag strategy for visual control, 268–293
    red tags and signboards, 265–268
    role in changeover improvement, 533
    seiketsu (cleaned up), 246–247
    seiri (proper arrangement), 243–245
    seiso (cleanliness), 246
    seiton (orderliness), 245–246
    shitsuke (discipline), 247–249
    signboard strategy for visual orderliness,
        293–306
    visible 5Ss, 249–262
5S badges, 255, 257

5S checklists, 258, 259
    for changeover, 818–819
5S contests, 258
5S implementation memo, case study, 286
5S maps, 261–262
5S memos, 755–757
5S mini motto boards, 255, 257
5S patrol score sheet, 258–259, 260
5S photo exhibit, 260
5S radar chart, 754
5S stickers, 257, 258
5S-related forms, 747
    5S checklists for factories, 747–749
    5S checklists for offices, 753
    5S checklists for workshops, 750–752
    5S memos, 755–757
    5S radar chart, 764
    cleaning checklist, 768–770
    display boards, 775–776
    five-point checklist to assess cleaned-up
        status, 771–774
    lists of unneeded inventory and equipment,
        764–767
    red tag campaign reports, 761–763
    red tags, 758–760
5W1H Sheet, 131, 744–746
    and on-site experience, 233, 235
    first Why guidelines, 233
    follow-up after line stops, 234
    three 5W1H essentials, 233
    waste prevention with, 232–233
7 Ms plus E&I, 551, 552

A

A-B control, 676, 677
Acceptable Quality Level (AQL), 121, 122

Accident-prevention devices, 698
    poka-yoke, 699–709
Accidents
    plywood gluing process, 696
    reasons for, 685–687
Actual work environment. *See* On-site
        experience
Added-value work, 75
Address signboards, 299
Adjustment errors, 560
Adjustment waste, 510
Administrative waste, 173
    and clerical standardization, 229
    disposal case study, 291
After-sales service part requests, 162
Air-processing machines, waste in, 156–157
Airplane *andon,* 466
Alerts, 672
Aluminum casting deburring operation,
        operations analysis table, 192
Amplifier-equipped proximity switches, 578
Andon systems, xiii, 11, 129, 231, 676, 679,
        680, 682
    hire method for using, 465–466
    illuminating factory problems with, 464
    operation *andon,* 468–469
    paging *andon,* 465–466
    progress *andon,* 469–470
    types of, 465
    warning *andon,* 466–468
    waste prevention using, 232
Anticipatory buying, 162
Anticipatory large lot production, 286–287
Anticipatory manufacturing, 162
Apparent minor defects, 680
Appropriate inventory, 96
Arm motions, 220–221
Arrow diagrams, 187–188, 211, 347, 730
    applications, 730
    examples, 731–732
    printed circuit board assembly shop, 189
    tutorial, 187–190
ASEAN countries, xi
Assembly line
    applying *jidoka* to, 660
    extending *jidoka* to, 676–682
    jidoka o prevent oversights in parts
        assembly, 680–681
    stopping at preset position, 69, 678–680
Assembly method error, 678
Assembly parts, exchange of, 499

Assembly processes
    changeover example, 495
    changing to meet client needs, 20
    establishing specialized lines for, 371–373
    kanban in, 447–448, 448
    management of, 81
    manpower reduction example, 428
    multi-process operations in, 363
    standing while working in, 355–359
    warning *andon* for long, 468
    warning *andon* for short, 467
Assembly step omission, 592
Attitude adjustment, 143–144
Auditory control, 120, 231
    waste prevention with, 230–232
Auto feed time, 635
Auto parts machining line, 400
Auto-extract devices, 657
Auto-input devices, 657
Automatic shut-off, 672
Automation, 102–103, 111
    limitations of, 79
    reinforcement of waste by, 111
    vs. Jidoka (human automation), 656,
        657–658
Automobile assembly plant, parts shelves,
        460, 461
Awareness revolution, 103, 104, 105, 159,
        176, 199, 344, 641, 721
    as premise for JIT production, 46, 344
    as prerequisite for factory improvement,
        13–15

B

Back-door approach, to waste discovery,
        181–183
Back-to-the-source inspection, 168, 170–172
Backsliding, 229
Basic Spirit principles, 203, 204
Baton touch zone method, 359, 368, 491, 492
Bills of materials, 81, 83
Blade exchange, 498
Board insertion errors, 594
Body movement principles, 220–221, 220–223
Body, as main perceptive instrument, 134
Bolt removal, eliminating need for, 521, 536
Bolt tightening reductions, 520
Boltless approach, 535

Boltless die exchange, 523
Bolts
    as enemies, 509, 535
    making shorter, 535
Bottlenecked processes, 364
Bottom-up improvements, 134–139
Bracket attachment errors, 603
Brainstorming, 208
    factory problems as opportunities for, 208
Breakdowns
    for standard operations charts, 638
    reducing through 5Ss, 241
Bridge defects, 598
Brush omission errors, 609
Buyer's market, 18
Bypass method, as leveling technique,
    491–492

C

Capacity imbalances, 161–162
    between processes, 214–215
    overcoming through 5Ss, 239
    retention and, 161–162
Capacity leveling, 21
Capacity requirements planning (CRP), 442
Capacity utilization rates, 68, 331, 341, 684
    and variety of product models, 504
Capacity-load imbalances, 151
Capital procurement, 93
Caravan style operations, 407, 423
Case studies
    drilling machine worker separation,
      669–672
    factory revolution, 287–289
    red tag strategy at Company S, 285–289
Cash-convertible assets, 93
Caster strategy, 349–350, 420. *See also*
      Movable machines
Chair-free operations, 19
Change, resistance to, 40, 201
Changeover 5S checklist, 512
Changeover costs, 73
    component costs, 73, 74
    variation in, 597
Changeover improvement list, 505, 810–811
    time graph analysis, 513

Changeover improvement procedures,
    500–502
    applying 5Ss to eliminate waste, 502
    changeover improvement list, 505
    changeover *kaizen* teams for, 503–506
    changeover operations analysis, 501–502,
      506–508
    changeover operations analysis charge,
      508
    changeover results table, 507
    eliminating waste with 5Ss, 508–511
    external changeover procedures, 501
    identifying wasteful operations, 508–511
    improving external changeover, 502
    improving internal changeover, 502
    injection molding process case study,
      515–517
    internal changeover procedures, 500
    kaizen team, 501
    public changeover timetable, 505
    transforming internal changeover to
      external changeover, 502
    waste, 501
Changeover improvement rules, 532–533
    role of 5Ss, 533–534
Changeover *kaizen* teams, 501, 503–506
Changeover operations, 71, 347, 723
    adjustment waste in, 510
    and introduction of synchronization, 373
    approach to changeover times, 499–500
    assembly line improvement example, 495
    avoidance of, and retention, 162
    balancing costs with inventory
      maintenance costs, 72
    changing standard parameters, 499
    exchange of dies and blades, 498
    exchanging assembly parts, 499
    external changeover time, 500
    general set-up, 499
    in JIT production system, 11
    internal changeover time, 500
    minimizing number, 216
    procedures for improvement, 500–532
    production leveling strategies for, 494–495
    rationale for improvement, 497–498
    reducing through 5Ss, 242
    replacement waste in, 509–510
    seven rules for improving, 532–539
    shortening time for, 494
    standardizing, 538–539
    time-consuming nature of, 216, 219

types of, 498–499
within cycle time, 514
Changeover operations analysis, 501–502, 506–508, 535
chart, 508
Changeover results table, 507, 815–817
Changeover standards, standardizing, 537
Changeover times, 499–500
Changeover work procedure analysis charts, 812–814
Checking, 691
Cleaned up checklist, detail, 256
Cleaned up, visibly, 253
Cleaning checklist, 768–770
Cleanliness, 16, 246, 690–691
five-point checklist, 772
of machinery, 119
visible, 253
Cleanliness check cards, 692
Cleanliness control board, 691
Cleanliness inspection checklist, 254, 690, 692
Cleanliness, Checking, and Oiling (CCO), 689–693
training in, 708
Cleanup, 16, 246–247
Cleanup waste, in external changeover procedures, 511
Clerical standardization, 229
Client needs, as determinant of capacity, 22
Client orders, as basis for cycle time/pitch, 70
Color coding, 253
for maintenance, 693
for oil containers, 319
in changeover improvements, 534
in *kaizen* boards, 462
Color mark sensors, 574, 580
applications, 582
Combination charts, 224
clarifying human work *vs.* machine work with, 664
for standard operations, 223–226
steps in creating, 630–632
wood products manufacturer example, 226, 227
Communication
about 5S approach, 263
errors in, and defects, 555–556, 558
Compact equipment, 19, 117–118, 340–341, 427, 484
as condition for flow production, 340–341, 342

building flexibility through, 419
compact shotblaster, 354
compact washing unit, 356
cost savings from, 354
diecast factory case study, 375–376, 377
for multi-process operations, 398–399
separating human and machine work with, 431
Company cop-out, 107, 108
Company-wide efficiency, 68
Company-wide involvement, with 5S approach, 262
Complexity
and waste, 648
in moving parts, 694
Component efficiency, 66
Computer-based management, 81
Computerization
and waste, 83
expendable material created by, 157
waste-making, 81
Computers
failure to shorten physical lead-time, 5
red tagging, 278–281
Confirmed production schedule, 439
Constant demand, products *vs.* parts, 475–476
Contact devices, 570
differential transformers, 572
limit switches, 570
microswitches, 570
Container organization, for deliveries, 385
Continuous flow production time, 19
Continuous improvement, 211
Control devices, 567
Control standardization, 228
Control/management waste, 149
Conveyance liveliness index, 304
Conveyance waste, 69, 149, 163–166, 173, 176, 180, 187, 336, 355–356, 392
links to retention, 164
Conveyor systems
appropriate use of, 70–71
improving equipment layout to eliminate, 79
waste hidden in, 67
Conveyor use index, 137
Conveyor waste, 155–156
Cooperative operation confirmation chart, 788–790
Cooperative operations, 367–371, 419
improvement steps for, 369
labor cost reduction through, 427–430

placing parts in front of workers for, 370
VCR assembly line example, 429
Cooperative operations zones, 370–371
Coordinated work, waste in, 67
Corporate balance sheet, inventory in, 94
Corporate culture, 15
Corporate survival, xii
Corrective maintenance, 688
Cost reduction, 69–71
   and effort invested, 71–74
   and profit, 36
   resistance arguments, 200–201
   through 5Ss, 239
   through *jidoka,* 659
Cost, in PQCDS approach, 3
Cost-up method, 35
Countable products, 119
Craft unions, *vs.* enterprise unions, 393–394
Crane operations, safety *poka-yoke,* 706
Cube improvements, 27
Current assets, 93
Current conditions, analysis to discover
      waste, 185–198
Current liabilities, 94
Current operating conditions, 24
Customer complaints, *vs.* defects, 547–548
Customer lead-time, 99
Customer needs, loss of concern for, 113–114
Customers, role in efficiency improvement,
      62–65
Cutting tools
   layout, 317
   orderliness applied to, 316–319
   placement, 317
   storage, 318
   types of, 317
Cycle list method, 487–489
   reserved seats and, 489–490
Cycle tables, 485
Cycle time, 19, 22, 332, 337, 363, 433, 630,
      634, 637, 647. *See also* Pitch
   and production leveling, 421–422
   and standard operations, 625
   as leveling technique, 485–487
   calculating, 487
   completing operations within, 636
   factors determining, 70
   for standard operations charts, 637
   overproduction and, 677
   smaller equipment for maintaining, 398
   vs. speed, 116

D

Deburring omissions, 589
Defect identification, 546
   and causes of defects, 558–561
   and factors behind defects, 550–558
   defects as people-made catastrophes,
      546–547
   inspection misunderstandings, 547–550
Defect prevention, 168, 177
   assembly step omission, 592
   board insertion errors, 594
   bracket attachment errors, 603
   bridge defects, 598
   brush omission errors, 609
   deburring omissions, 589
   defective-nondefective part mixing errors,
      613
   drilling defects, 600, 675–676
   E-ring omission errors, 611
   equipment improvements for, 640
   gear assembly errors, 614
   grinding process omission, 591
   hole count errors, 588
   hole drilling omission, 593
   hose cut length variations, 597
   incorrect drill position, 601
   left-right attachment errors, 615
   mold burr defects, 674–675
   nameplate omission errors, 608
   packing omission errors, 610
   part omission errors, 607
   pin dimension errors, 595
   press die alignment errors, 596
   product set-up errors, 602
   spindle hole punch process omission, 590
   tap processing errors, 606
   tapping operations, 673–674
   through 5Ss, 241
   through automatic machine detection, 403
   through *jidoka*
   through simplified production operations,
      549
   torque tightening errors, 599
   with *kanban,* 441–442
   with multi-process operations, 392
   workpiece direction errors, 605
   workpiece positioning errors, 605
   wrong part assembly errors, 612
Defect production waste, 176–177, 180

Defect reduction, 168, 544
   with compact machinery, 399
Defect signals, 567
Defect-prevention devices, 659, 669, 673
Defective assembly parts, 678
Defective item display, 457, 458
Defective products
   and inventory, 92
   counting, 119
   ending downstream processing of,
      544–545
   factories shipping, 542
   increases with shish-kabob production,
      158
   increasing inspectors to avoid, 542–544
   inventory and, 90–91
   noncreation of, 545–546
   waste in making, 159
Defective/nondefective part mixing errors,
   613
Defects
   and communication errors, 555–556, 557,
      558
   and inspection, 548
   and production method errors, 555, 557
   and surplus products, 549
   as human-caused catastrophes, 546–547
   causes, 558–561
   due to human errors, 551, 553, 557, 558
   due to machine errors, 554–555, 557
   factors behind, 550–558
   in materials, 553–554, 557
   relationship with errors and inspection,
      543
   stoppages for, 567
   ten worst causes, 561
   vs. customer complaints, 547–548
Delays, reducing through 5Ss, 242
Delivery
   and loading methods, 379
   and transport routes, 380–382
   and visible organization of containers, 385
   applying flow concept to, 378–382
   color coding strategy, 384
   FIFO strategy, 384–385
   frequency of, 380
   in PQCDS approach, 3
   self-management by delivery companies,
      383
Delivery company evaluation table, 382,
      791–793
Delivery schedules, shortening of, 2

Delivery sites
   applying flow concepts to, 382–385
   establishment of, 383
   product-specific, 384
Detach movement, automation of, 671–672,
      673
Deterioration, 686
   and accidents, 685
   preventive measures, 688
   reversing, 688
Die exchange, 498
   improvement for boltless, 523
   minimizing, 497
Die height standardization, 526–527
Die storage sites, proper arrangement and
      orderliness applied to, 530–531
Diecast deburring line, 351
Diecast factory, flow production case study,
      373–378
Differential transformers, 572
Dimensional tolerances, 686
Dimensions, enlarging, 311
Disaster prevention measures, waste in, 159
Discipline, 16, 247–249
   JIT Improvements as, 130
   visible, 254–255
Displacement sensors, 574
   applications, 579–580
Display boards, 775–776
Distribution, applying JIT to, 47
Diversification, 2, 117, 415, 416
   of consumer needs, 62
   through 5Ss, 242
Do it now attitude, 236
Doing, as heart of JIT improvement, 133
Dot it now attitude, 236
Double-feed sensors, 576
   applications, 584
Downstream process control inspection
      method, 169, 170
Drill bit replacement, external changeover
      improvement, 532, 533
Drill bit storage method, improvements, 235
Drill operation, before improvement, 670
Drill position errors, 601
Drilling defects, 600
   avoiding downstream passing of, 675–676
Drilling machine, 662
   detach movement, 671–672
   hold motion automation, 671
   jidoka case study, 669–672

safety plate for, 703, 704
separating human from machine work on, 402

E

E-ring omission errors, 611
Economical lot sizes, 72
Economy of motion, 642
Economy of scale, 45
Efficiency
  and production leveling, 69
  approaches to, 59–61
  customer as driver of, 62
  estimated *vs.* true, 59–61
  individual and overall, 66–69
  maximizing at specific processes, 484
  overall, 484, 492
  raising in individual processes, 68
  shish-kabob *vs.* level production
    approaches, 484, 486
Electric screwdrivers, combining, 315
Emergency *andon*, 464
Employees, as basic asset, 108
End-of-month rush, 162
Energy waste
  due to inventory, 325
  through inventory, 91
Engineering technologies, applying JIT
  improvement to, 334
Engineering-related forms, 777
  5S checklist for changeover, 818–819
  changeover improvement lists, 810–811
  changeover results tables, 815–817
  changeover work procedure analysis
    charts, 812–814
  cooperative operation confirmation chart,
    788–790
  delivery company evaluation charts,
    791–793
  JIT delivery efficiency list, 794–796
  line balance analysis charts, 785–787
  model and operating rate trend charts,
    805–807
  multiple skills evaluation chart, 799–801
  multiple skills training schedule, 797–798
  P-Q analysis lists/charts, 777–781
  parts-production capacity work table,
    822–824

poka-yoke/zero defects checklist,
    820–821
  process route diagrams, 782
  production management boards, 802–804
  public changeover timetables, 808–809
  standard operations combination chart,
    825–826
  standard operations form, 831–833
  summary table of standard operations,
    827–828
  work methods table, 829–830
Enterprise unions, *vs.* craft unionis, 393–394
Enthusiasm, as prerequisite for innovation,
    143, 144
Equal-sign manufacturing cells, 362
Equipment
  applying *jidoka* to, 660
  automation and human automation,
    102–103
  compact, 19, 117–118
  ease of maintenance, 119
  ease of operation, 118
  ergonomics recommendations, 222
  for flow production, 389
  improvements facilitating standard
    operations, 640
  modification for multi-process operations,
    406
  movability, 64–65, 117–118
  obtaining information from, 119–120
  shish-kabob *vs.* flow production
    approaches, 331
  standardization in Japanese factories, 395
  versatility and specialization, 116–117
  vs. work operations improvements,
    103–108
Equipment breakdown, 708
  acceptance of, 683
  apparent minor defects, 680
  below-expectation performance, 686
  breakdown stage, 686
  intermittent stoppage stage, 686
  latent minor defects stage, 680
  preventing, 693–695
  stages, 685, 687
Equipment constitution, 694
Equipment costs
  and *jidoka*, 666
  vs. labor costs, 658
Equipment improvement, 103, 104, 106
  and company cop-out, 108
  based on manufacturing flow, 114–120

cost of, 104, 109–111
irreversibility of, 112, 113–114
not spending money on, 207–208
reinforcement of waste by, 111–112
twelve conditions for, 114–120
typical problems, 108–114
Equipment improvement problems, 110
Equipment layout
applying *jidoka* to, 662
as condition for flow production, 336–337, 342
for flow production, 389
in order of processing, 353–355
shish-kabob *vs.* flow production approaches, 330
Equipment signboards, 295
Equipment simplification, 400
Equipment waste, 149
Error control, 567
Error prevention boards, 457, 458
Errors, relationships with defects and inspection, 543
Estimate-based leveling, 23
Estimated efficiency, 59–61
Estimated lead-time, 98–99
Estimated production schedule, 439
Estimated quality, 122
Excess capacity, 174
Excuses, 202, 205
Expensive improvements, failure of, 206
Experiential wisdom, 210–211
External changeover improvements, 529–532
carts reserved for changeover, 531–532
drill bit replacement example, 532
proper arrangement and orderliness in die storage sites, 530–531
External changeover procedures, 501
cleanup waste in, 511
improving, 502
preparation waste in, 510
waste in, 510–511
External changeover time, 500

F

Factory
as best teacher of improvements, 134–139
as living organism, 230
Factory bath, 270

Factory graveyards, 73
Factory improvement
5Ss for, 15–17
awareness revolution prerequisite, 13–15
shortening physical lead-times through, 6
vs. JIT improvements, 13
Factory layout diagram, 188
Factory myths
anti-JIT production arguments, 40–44
fixed ideas and JIT production approach, 44–47
sales price/cost/profit relations, 35–40
Factory problems, 326
as brainstorming opportunity, 208
illuminating with *andon,* 464
stopgap responses to, 150
ubiquitousness of, 251
Factory revolution, 287–289
Factory-based innovation, xiii, 133
Factory-wide efficiency, 68
Feed motion, 664
applying *jidoka* to, 665
jidoka, 670, 671
Feet, effective use of, 221–222, 223
Fiber optic switches, 575, 579
Finance, inventory and, 92–95
Fine-tuning waste, 537
removal, 523–527
Fingernail clipping debris, device preventing, 247
First-in/First-Out (FIFO), 302–303, 461, 462
as delivery strategy, 384–385
Five levels of quality assurance achievement, 542–546
Five whys, 24, 130–134, 183, 184, 210, 236
applying to changeover improvements, 535
waste discovery through, 208–210
Five-point checklist, 771
for cleanliness, 772
for proper arrangement, 772
Five-point cleaned up checklist, 255, 257–258, 773, 774
Fixed ideas, 235
about conveyors, 156
avoiding for waste prevention, 235–236
direct challenge to, 43
eliminating for waste removal, 204
kanban, 447
large lot production, 417
wall of, 210
Fixed liabilities, 94

Flexibility
  in baton touch zone method, 491
  mental origins of, 420
Flexible production, 419
Flexible staff assignment system, 63, 65, 417, 419
Flow analysis, 188
  summary chart, 189, 190
Flow components, 56
Flow control, 567
Flow devices, 108, 109
Flow manufacturing, xii, 9–10, 49, 64, 70, 79–84. *See also* One-piece flow
  and line improvements, 25
  making waste visible by, 17
  role in JIT introduction, 17–19
  seven requirements, 19
Flow of goods, 159–160, 641, 646
  device improvements facilitating, 638–640
Flow production, 50, 321, 564–565
  and evils of inventory, 324–328
  and inventory accumulation, 321–324
  applying to delivery sites, 382–385
  approach to processing, 329–330
  at diecast factory, 374, 376
  between factories, 332–333, 378–385
  caster strategy, 349–350
  defect prevention with, 721
  diecast factory case study, 373–378
  eight conditions for, 333–341
  equipment approach, 331
  equipment layout in, 330
  for production leveling, 492–494
  in medical equipment industry, 423
  in multi-process operations, 388
  in-process inventory approach, 331
  interrelationship of factors, 343
  lead time approach, 331
  operator approaches, 330–331
  preparation for, 344–350
  procedure for, 350–373
  rational production approach in, 330
  reducing labor cost through, 422–424
  sink cabinet factory example, 493
  steps in introducing, 343–373
  straight-line method, 340
  U-shaped manufacturing cell method, 340
  vs. shish-kabob production, 328–332
  waste elimination techniques, 341–342
  within-factory, 332–333, 333–341
Flow shop layout, 395
Flow unit improvement, 639

Forms, 711–714
  5S-related, 747–776
  engineering-related, 777–833
  for standard operations, 626–628
  JIT introduction-related, 834–850
  overall management, 716–729
  waste-related, 730–746
Free-floating assembly line, 356, 357
Full lot inspection, 120–122
Full parallel operations, 225
Full work system, 175, 365, 676–677
  A-B control, 677
  devices enabling, 368
  pull production using, 367
Function-specific inventory management, 305

G

Gear assembly errors, 614
General flow analysis charts, 733–734
General purpose machines, 331, 340
Golf ball *kanban* systems, 450–451
Graph time, 633
Gravity, *vs.* muscle power, 221
Grinding process omission, 591
Groove processing lifter, separating human/machine work, 649
Group Technology (GT) lines, 347
  for line balancing, 491

H

Hand delivery, 365
Hand-transferred one-piece flow, 337, 338
  pull production using, 366
Handles/knobs, 223
Hands-on improvements, 9, 140
Height adjustments, avoiding, 538
Hirano, Hiroyuki, xiii
Hold motion, automation of, 671
Hole count errors, 588
Hole drilling omission, 593
Horizontal development, 24–25, 391
Hose cut length variations, 597
Household electronics assembly, labor cost reduction example, 428

Human automation, 12, 62, 102–103, 159,
554, 655. *See also Jidoka* (human
automation)
and removal of processed workpieces,
668
and setup of unprocessed
workpieces/startup, 669
applying to feeding workpieces, 665
applying to return to starting positions,
667
for multi-process operations, 402
Human error waste, 173, 674
and defect prevention, 551–553
basic training to prevent, 562–563
defects and, 546–547
eliminating by multiple skills training, 563
minimizing, 177
Human movement
body movement principles, 220–223
removing wasteful, 217–223
Human work, 658
clarifying with combination charts, 664
compact PCB washer example, 431
procedure for separating from machines,
682–689
separating from machine work, 64,
118, 400–402, 406, 430–432, 640,
649–650, 660–662, 702, 703
Humanity, coexistence with productivity,
387–388

I

Idle time waste, 66, 67, 69, 156, 173, 178–179,
180, 682
cooperative operations as solution to,
367–371
Impact wrench, 680, 681
Implementation, 139–144
of multi-process operations, 405
Implementation rate, for waste removal,
205–206
Improvement
and enthusiasm, 143, 144
intensive, 266–268
making immediate, 538
poor man's approach, 106
spending on, 284

spirit of, 43
with visual control systems, 453–454
Improvement days, weekly, 32
Improvement goals, 191
Improvement lists, 33–34
Improvement meetings, 32–33, 33
Improvement promotion office, 31–32
Improvement results chart, 462, 844–845
Improvement teams, 31, 32
Improvements
bottom-up *vs.* top-down, 134–139
factory as best teacher, 134–139
implementing, 24
mental *vs.* physical, 130–134
passion for, 143–144
promoting, 126–130
pseudo, 126–130
Improving actions, 220
In Time concept, 48
In-factory *kanban,* 443, 444–445
In-line layout, 364, 376
compact shotblaster for, 377
washing units, 365
In-process inventory, 101, 102, 161, 175, 447,
484
and standard operations, 625–626
for standard operations charts, 637
production *kanban* for, 445
reduction of, 647, 649
relationship to *kanban,* 435
shish-kabob *vs.* flow production
approaches, 331
symbols for standard operations charts,
637
Inconsistency, 152, 643
eliminating, 151
Independent improvement, 688–689
Independent maintenance, 688–689
Independent process production, 53
inflexibility in, 54
Independent quality control inspection
method, 169, 170
Individual efficiency, 66–69
Industrial engineering (IE), xii
and conveyor use index, 137
motion study in, 642
vs. JIT method, 136
Industrial fundamentalism, 105, 106
Industrial robots, 668
Inexpensive machines, versatility of, 117
Information inspection, 168, 169
Inherent waste, 79–84

Injection molding process
    burr defect prevention, 674
    internal changeover improvement case
        study, 515–517
Injuries
    reasons for, 695–697
    reducing through 5Ss, 241
Innovation, 13, 37
    and JIT production, 47–49
    enthusiasm as prerequisite for, 143
    factory-based, xiii
    in JIT production, 47–49
    JIT production as, 27
Inspection, 56, 160, 187
    back-to-the-source inspection, 170–172
    eliminating need through *jidoka,* 674
    failure to add value, 168
    failure to eliminate defects, 120
    increasing to avoid defective products,
        542–544
    information inspection, 169
    preventive, 564
    relationship to defects, 543, 547–550
    sorting inspection, 169
Inspection buzzers, waste prevention with,
    232
Inspection functions
    building into JIT system, 119
    full lot inspection, 120–122
    sampling inspection, 120–122
Inspection waste, 149
Inspection-related waste, 167–168
Integrated tool functions, 223
Intensive improvement, 266–268
    timing, 268
Interest payment burden, 324, 326
    inventory and, 90
Intermittent stoppage stage, in equipment
    breakdown, 686
Internal changeover improvements, 518,
    534–535
    bolt tightening reductions, 520
    boltless die exchange, 523
    die height standardization, 526–527
    eliminating need to remove bolts, 521
    eliminating nuts and washers, 521
    eliminating replacement waste, 518–523
    eliminating serial operations, 527–529
    establishing parallel operations, 528
    one-touch tool bit exchange, 522
    protruding jigs *vs.* manual position
        setting, 524

removing fine-tuning waste, 523–527
    spacer blocks and need for manual dial
        positioning, 526
    spacer blocks and need for manual
        positioning, 524–525
    tool elimination, 519–520
Internal changeover procedures
    changing to external changeover, 511–518,
        534
    improving, 500, 502
    PCB assembly plant case study, 513–514
    transforming to external, 502
    turning into external changeover, 511–518
    waste in, 509–510
    wire harness molding process case study,
        517–518
Internal changeover time, 500
Inventory
    advance procurement requirements, 325
    and conveyance needs, 90
    and defects, 90–91, 92
    and energy waste, 91
    and finance, 92–95
    and interest-payment burden, 90
    and lead-time, 87–89, 88
    and losses due to hoarded surpluses, 325
    and materials/parts stocks, 91
    and price cutting losses, 325
    and ROI, 95
    and unnecessary management costs, 91
    and waste, 48
    as cause of wasteful operations, 325
    as evasion of problems, 176
    as false buffer, 95, 101
    as JIT consultant's best teacher, 89
    as opium of factory, 92–95
    as poor investment, 95–98
    breakdown by type, 161
    concealment of factory problems by, 91,
        92, 326, 327
    evasion of problems with, 163
    evils of, 90–92, 324–328
    FIFO storage method, 303
    in corporate balance sheet, 94
    incursion of maintenance costs by, 325
    interest payment burden due to, 324
    management requirements, 325
    product, in-process, materials, 101, 102
    red tagging, 281–282
    reducing with once-a-day production
        scheduling, 480–481

shish-kabob *vs.* level production approaches, 484–485
space waste through, 90, 325
unbalanced, 161
wasteful energy consumption due to, 325
with shish-kabob production, 158
zero-based, 98–102
Inventory accumulation
and caravan operations, 322
and changeover resistance, 322
and distribution waits, 322
and end-of-month rushes, 323
and faulty production scheduling, 323
and just-in-case inventory, 323
and obsolete inventory flow, 321
and operator delays, 322
and resistance to change, 322
and seasonal adjustments, 323–324
and standards revision, 323
and unbalanced capacity, 322
multiple-process sources of, 322
reasons for, 321
Inventory assets, 715
Inventory control, 126
Inventory flow, obsolete, 321
Inventory graveyard, 324
Inventory liveliness index, 303–304
Inventory maintenance costs, 72
Inventory management
function-specific method, 305
product-specific method, 305
with *kanban*, 436
Inventory reduction, 87, 89, 125
case study, 288, 289, 377
Inventory stacks, 303
Inventory waste, 175–176, 180
Irrationality, 152, 643
eliminating, 151
Item characteristics method, 568, 569
Item names, for signboards, 299–300
Ivory tower syndrome, 22

J

Japanese industrial structure, 1980s transformation of, xi
Jidoka (human automation), 12, 62, 102–103, 103–108, 655, 724
applying to feeding workpieces, 665
automation *vs.*, 656, 657–658
cost considerations, 667, 669
defect prevention through, 672–676
detach movement, 671–672
drilling machine case study, 669–672
extension to assembly line, 676–682
feed motion, 670
full work system, 676–677
manual labor *vs.*, 655, 656
mechanization *vs.*, 656
preventing oversights in nameplate attachments, 681–682
steps toward, 655–657
three functions, 658–660
Jigs
5-point check for orderliness, 256
applying orderliness to, 307
color-coded orderliness, 368–369
combining, 314
easy-to-maintain orderliness for, 307
eliminating through orderliness strategies, 313–316
indicators for, 308
outlined orderliness, 309
JIT delivery efficiency list, 794–796
JIT improvement cycle, 144
roles of visual control tools in, 473
JIT improvement items, 837–840
JIT improvement memo, 836
JIT improvements, 12, 13
"doing" as heart of, 133
and changeover costs, 74
and parts list depth, 82
as discipline, 130
as religion, 138
as top-down improvement method, 135
basis in ideals, 12
case study, 288
cube improvements, 27
factory as true location of, 34
from within, 139–143
hostile environment in U.S. and Europe, 107
improvement lists, 33–34
improvement meetings, 32–33
improvement promotion office, 31–32
lack of faith in, 41
line improvements, 25–26
plane improvements, 26–27
point improvements, 25
promoting and carrying out, 30–34
requirement of faith, 139

sequence for introducing, 21
seven stages in acceptance of, 140–144
ten arguments against, 299
vs. JIT production management, 7
vs. labor intensification, 86
weekly improvement days for, 32
JIT innovation, 13
JIT introduction steps, 12–13
5Ss for factory improvement, 15–17
awareness revolution step, 13–15
department chiefs' duties, 28–29, 30
division chiefs' duties, 28
equipment operators" duties, 30
factory superintendents' duties, 28–29
flow manufacturing, 17–19
foremens' duties, 30
leveling, 20–22
president's duties, 28
section chiefs' duties, 30
standard operations, 23–24
JIT introduction-related forms, 834
improvement memo, 836
improvement results chart, 844–845
JIT leader's report, 849–850
JIT Ten Commandments, 834–835
list of JIT improvement items, 837–840
weekly report on JIT improvements,
846–848
JIT leader's report, 849–850
JIT Management Diagnostic List, 715–718
JIT production
adopting external trappings of, 472
as new field of industrial engineering, xii
company-wide promotion, 28, 29
elimination of waste through, xi
five stages of, 719, 721, 726, 728
guidance, education and training in, 30
hands-on experience, 30
in-house seminar, 343
innovation in, 47–49
linked technologies in, 334
promotional organization, 31
radar chart, 727
setting goals for, 28
structure, 720
JIT production management
distinguishing from JIT improvements, 7
vs. conventional production management,
1–3
JIT production system
as total elimination of waste, 145
changeover, 11

flow manufacturing, 9–10
from vertical to horizontal development,
24–27
human automation, 12
introduction procedure, 12–14
jidoka, 12
kanban system, 10
leveling, 11
maintenance and safety, 12
manpower reduction, 10
multi-process handling, 10
organizing for introduction of, 27–30
overview, 7–9
quality assurance, 11
standard operations, 11–12
steps in establishing, 14
view of waste, 152
visual control, 10–11
JIT radar charts, 719, 727, 729
JIT study groups, 15
JIT Ten Commandments, 834–835
Job shop layout, 395
Just-in-case inventory, 323
Just-In-Time
anatomy of, 8–9
and cost reduction, 69–71
as consciousness improvement, 139–143
functions and five stages of development,
728
innovation and, 47–49
view of inspection work, 168

K

Kaizen boards, 462
visual control and, 471–473
with improvement results displays, 463
Kanban systems, xii, xiii, 7, 8, 10, 11, 52,
54, 174, 231, 365, 692, 722
administration, 447–451
and defect prevention, 441–442
and downstream process flow, 441
and in-process inventory, 435
applying to oiling, 693
appropriate use of, 70–71
as autonomic nervous system for JIT
production, 440
as tool for promoting improvements, 441
as workshop indicators, 442

differences from conventional systems, 435–437
factory improvements through, 440–441
fixed ideas about, 447
functions, 440–441
in processing and assembly lines, 447–448
in-factory *kanban,* 444–445
novel types, 450–451
production *kanban,* 445
production leveling through, 442
purchasing-related, 449–450
quantity required, 445–447
rules, 441–442
signal *kanban,* 445
supplier *kanban,* 443
types of, 442–447
visual control with, 457
vs. conventional production work orders, 437–439
vs. reordering point method, 435–437
waste prevention with, 232

L

L-shaped line production, 360
Labor cost reduction, 415, 418, 722
    and elimination of processing islands, 421
    and mental flexibility, 420
    and movable equipment, 420–421
    and multi-process operations, 421
    and production leveling, 421–422
    and standardized equipment and operations, 421
    approach to, 415–418
    display board for, 433–434
    flow production for, 422–424
    multi-process operations for, 424–426
    multiple skills training schedule for, 432–433
    steps, 419–422
    strategies for achieving, 422–432
    through cooperative operations, 427–430
    through group work, 426–427
    through separating human and machine work, 430–432
    visible, 432–434
    vs. labor reduction, 417–418
Labor cost reduction display board, 433–434

Labor intensity/density, 84–86
    vs. production output, 86
Labor per unit, 649
Labor reduction, 63, 418, 647
    vs. labor cost reduction, 417–418
    vs. worker hour minimization, 66–69
Labor savings, 418
Labor unions, 107. *See also* Craft unions; Enterprise unions
    and multi-process operations, 393–394
Labor-intensive assembly processes, 217
Large lot sizes, 18, 62, 73, 278, 321, 398, 483, 598
    and changeover times, 216
    and machine waste, 155
    as basis of production schedules, 476
    case study, 286–287
    fixed ideas about, 417
    switching to small-lot flow from, 639
Large machines waste, 154–155, 331
Large-scale container deliveries, 381
Latent minor defects, 680
Latent waste, 198
Lateral development, 27, 378, 505, 506
Lateral improvement makers, 167
Lathes, 682
    three kinds of motion, 663
    worker separation from, 702
Layout improvement, 638
Lead-time
    and inventory, 88
    and lot sizes, 498
    and production lot size, 72
    and work stoppage, 59–61
    estimated *vs.* real, 98–99
    inventory and, 87–89
    lengthened with shish-kabob production, 158
    paper, 4, 5
    physical, 5
    product, 4
    reduction with multi-process operations, 393
    shish-kabob *vs.* flow production approaches, 331, 486
    shish-kabob *vs.* level production approaches, 484–485
    shortening by reducing processing time, 55
Leadership, for multi-process operations, 404–405
Left-right attachment errors, 615

Leg motion, minimizing, 221
Level production, 475, 723. *See also* Leveling
   as market-in approach, 482
   vs. once-a-day production, 481
   vs. shish-kabob production, 482–485, 486
Leveling, 50, 476. *See also* Level production;
      Production leveling
   and production schedule strategies,
      477–482
   approach to, 476–477
   capacity and load, 21
   estimate-based, 23
   reality-based, 23
   role in JIT introduction, 20–22
   role in JIT production system, 11
   techniques, 482–492
Leveling techniques, 485
   baton touch zone method, 491
   bypass method, 491–492
   cycle list method, 487–489
   cycle tables, 485
   cycle time, 485–487
   nonreserved seat method, 487–489
   reserved seat method, 489–490
Limit switches, 403, 470, 570, 676, 677, 706,
      708
Line balance analysis charts, 785–787
Line balancing
   at PCB assembly plant, 514
   SOS system for, 217
   strategies for, 491
Line balancing analysis tables, 358
Line design, based on P-Q analysis, 346, 347
Line efficiency, 68
Line improvements, 25–26
Line stops, 470
   5W1H follow-up after, 234
   at preset positions, 678–680
   with *poka-yoke* devices, 675
Lined up inventory placement, 304–306
Linked technologies, in JIT production, 334
Litter-preventive device, for drill press, 248
Load leveling, 21
Loading methods, 379
Long-term storage, case study, 291
Lot sizes, 45, 87
   and lead time, 72
   large *vs.* small, 71–74
Lot waiting waste, 215–216, 219
   waste removal, 219
Low morale, 16

M

Machine errors
   and defect prevention, 554–555
   poka-yoke to prevent, 564
Machine operating status, *andon*
      notification of, 466
Machine placement, waste and, 185
Machine signboards, 295
Machine standardization, 228
Machine start-up, applying *jidoka* to, 663,
      668
Machine work
   clarifying with combination charts, 664
   compact PCB washer example, 431
   separating from human work, 64, 118,
      400–402, 406, 430–432, 640,
      649–650, 660–662
Machine/people waiting, 214
Machines
   as living things, 120–122
   shish-kabob *vs.* level production
      approaches, 484, 486
   with strong constitution, 708
Machining line, full work system, 677
Maintenance, 683, 725
   and accidents, 685–687
   and possible utilization rate, 684–685
   breakdown prevention, 693–695
   Cleanliness, Checking, and Oiling (CCO)
      approach, 689–693
   defined, 684–689
   existing conditions, 683–684
   full-fledged, 708–709
   improving through 5Ss, 241
   in JIT production system, 12
   of equipment, 119
Maintenance campaigns, 687–689
Maintenance errors, 560
Maintenance prevention, 688
Maintenance technicians, 689
Make-believe automation, 79
Man, material, machine, method, and
      management (5Ms), 152, 153
Management-related forms, 715
   five stages of JIT production, 719, 721–725
   JIT Management Diagnostic List, 715–718
   JIT radar charts, 719
Manpower flexibility, 338
Manpower needs, based on cycle time, 22

Manpower reduction, 10, 62–65, 63, 337, 392
  household electronics assembly line
    example, 428
  improving efficiency through, 61
  through flow production, 422–424
Manual dial positioning, eliminating with
    spacer blocks, 526
Manual labor, 655, 656
Manual operations, two-handed start/stop,
    220
Manual position setting, eliminating need for,
    524
Manual work time, 635
Manual-conveyance assembly lines,
    progress *andon* in, 469
Manufacturing
  as service industry, 1
  five essential elements, 553
  nine basic elements (7Ms plus E&I), 552
  purpose of, 1
Manufacturing flow, as basis for equipment
    improvements, 114–120
Manufacturing process, components, 56
Manufacturing waste, 149
Market demand fluctuations, unsuitability of
    kanban for, 436
Market price, as basis of sales price, 35
Market-in production, xii, 416, 555
  level production as, 482
Marshaling, 306
Mass production equipment, 216, 219
Material handling
  building flexibility into, 419
  minimizing, 176
  vs. conveyance, 164
Material handling costs, 159, 163
Material requirements planning (MRP), 52
Materials flow
  device improvements facilitating, 638–640
  standard operations improvements, 641
Materials inventory, 101, 102
Materials waiting, 215, 218
Materials waste, 157
Materials, and defect prevention, 553–554
Measuring tools
  orderliness for, 318
  types, 319
Mechanization, 656
Medical equipment manufacturing,
    manpower reduction example, 423
Meetings, waste in, 158

Mental improvements
  vs. implementation, 140
  vs. real improvements, 130–134
Metal passage sensors, 574
  applications, 581
Microswitch actuators, 571
Microswitches, 570, 674
Milling machine, safety *poka-yoke* for,
    705–706
Minimum labor cost, 62
Missing item errors, 587, 607–611, 678
Mistake-proofing, 119
Mistakes, correcting immediately, 207
Mixed loads, 379
Mixed-model flow production, 492
Mizusumashi (whirligig beetle), 465
Model and operating rate trend charts,
    805–807
Model lines, analyzing for flow production,
    348
Mold burr defects, prevention, 674–675
Monitoring, *vs.* managing, 123–126, 126–130
Motion
  and work, 74–79
  as waste, 76, 78, 79, 84
  costs incurred through, 77
  economy of, 642
  lathes and, 663
  vs. work, 657, 659
Motion study, 642
Motion waste, 639
  improvements with standard operations,
    639
Motor-driven chain, 694
Movable machines, 64–65, 65, 117–118, 165,
    354, 420
  and caster strategy, 349–350
  building flexibility through, 419
Movement
  as waste, 178
  improving operational efficiency, 642–649
  non-added value in, 190
Muda (waste), 643
Multi-process operations, 10, 19, 64, 330,
    359, 362–363, 387–388, 417, 722
  abolishing processing islands for, 396–398
  and labor unions, 393–394
  as condition for flow production, 337–338
  basis for pay raises in, 394
  compact equipment for, 398–399
  effective leadership for, 404–405
  equipment layout for, 389

equipment modification for, 406
factory-wide implementation, 405
human assets, 389
human automation for, 402–403
human work *vs.* machine work in, 400–402
in wood products factory, 425
key points, 395–404
labor cost reduction through, 424–426
multiple skills training for, 400
one-piece flow using, 338
operational procedures for, 389
poka-yoke for, 402–403
precautions, 404–406
promoting perseverance with, 406
questions from western workers, 393–395
safety priorities, 403–404, 406
simplified work procedures for, 404
standard operations improvements, 639
standing while working for, 399–400
training costs for, 394–395
training for, 421
training procedures, 407–413
transparent operations in, 405
U-shaped manufacturing cells for, 395–396
vs. horizontal multi-unit operations,
    388–393
Multi-process workers, 331
    as condition for flow production, 339
    at diecast factory, 377
Multi-skilled workers, 19, 390
    and standard operations, 650–651
    building flexibility through, 419
Multi-unit operations, 338, 391
    vs. multi-process operations, 388–393
Multi-unit process stations, 390
Multiple skills contests, 405
Multiple skills evaluation chart, 799–801
Multiple skills maps, 432
Multiple skills score sheet, 410, 432
Multiple skills training, 425, 651
    defect prevention with, 563
    for multi-process operations, 400
    schedule for, 432–434
Multiple skills training schedule, 797–798
Multiple-skills training, 407
    demonstration by workshop leaders, 412
    during overtime hours, 409
    five-level skills evaluation for, 408
    hands-on practice, 412
    importance of praise, 413
    in U-shaped manufacturing cells, 410
    schedule, 409

team building for, 408
trainer roles, 413
workshop leader roles, 411
Mura (inconsistency), 643
Muri (irrationality), 643
Mutual aid system, 65

N

Nameplate omission errors, 608
    preventing with *jidoka*, 681–682
Needed items, separating from unneeded
    items, 266
Net time, for standard operations charts, 637
Newly Industrialized Economic Societies
    (NIES), xi
Next process is your customer, 51, 54, 132
Non-value-added steps
    as waste, 147, 171
    in inspection, 170
    in retention, 163
Noncontact switches, 572
    color mark sensors, 574
    displacement sensors, 574
    double-feed sensors, 576
    metal passage sensors, 574
    outer diameter/width sensors, 574
    photoelectric switches, 572, 574
    positioning sensors, 574
    proximity switches, 574
    vibration switches, 574
Nondefective products, counting, 119
Nonreserved seat method, 487–489
Nonunion labor, 394
Nuts and washers, eliminating as internal
    changeover improvement, 521

O

Oil containers, color-coded orderliness, 319
Oil, orderliness for, 318–319
Oiling, 691–693
    kanban for, 693
On-site experience, 190
    and 5W1H method, 233, 235
    by supervisors, 230, 233, 235

Once-a-day production scheduling, 480–482
Once-a-month production scheduling, 478–479
Once-a-week production scheduling, 479–480
One how, 24, 128, 130–134, 183
One-piece flow, 19, 64, 115–116, 165, 185, 419, 639. *See also* Flow manufacturing
    as condition for flow production, 335–336
    discovering waste with, 183–185
    hand-transferred, 338
    in multi-process operations, 388
    maintaining to avoid creating waste, 351–353, 353
    revealing waste with, 350–351, 352
    switching to, under current conditions, 184
    using current equipment layout and procedures, 336
One-touch tool bit exchange, 522
Operation *andon,* 464, 468–469
Operation errors, 560
Operation management, 81
Operation method waiting, 215, 218
Operation methods, conditions for flow production, 342
Operation step method, 568, 569
Operation-related waste, 173, 178, 180
Operational combinations, 193
Operational device improvements, 640
Operational rules, standard operations improvements, 639–640
Operations analysis charts, 735–736
Operations analysis table, 190–192, 735, 736
    aluminum casting deburring operation example, 192
Operations balancing, 219
Operations improvements, 103, 104, 105, 217
Operations manuals, 405
Operations standardization, 228
Operations, improving point of, 220
Operators
    conditions for flow production, 342
    diecast factory case study, 377
    maintenance routines, 691
    reducing gaps between, 370
    shish-kabob *vs.* flow production approaches, 330–331
Opportunistic buying, 162
Optical displacement sensors, 578
Oral instructions, avoiding, 556
Order management, 81

Orderliness, 16, 157, 245–246, 510
    applied ti die storage sites, 530–531
    applying to jigs and tools, 307
    beyond signboards, 302–306
    color-coded, 319, 384
    conveyance liveliness index, 304
    easy-to-maintain, 307, 310–313
    eliminating tools and jigs with, 313–316
    for cutting tools, 316–319
    for measuring tools, 318
    for oil, 318–319
    four stages in evolution, 312
    habitual, 302
    inventory liveliness index, 303–304
    just-let-go principle, 313, 314
    lined up inventory placement, 304–306
    made visible through red tags and signboards, 265–268
    obstacles to, 17
    visible, 252–253
Outer diameter/width sensors, 574
    applications, 578
Outlined orderliness, for jigs and tools, 309–310
Outlining technique, waste prevention with, 231
Overall efficiency, 66
Overkill waste, 173
Overload prevention devices, 706
Overproduction waste, 69, 174–175, 180
    beyond cycle time, 677
    preventing with A-B control, 676–677
Overseas production shifts, xi

P

P-Q analysis, 188, 345–346
P-Q analysis lists/charts, 777–781
Packing omission errors, 610
Paging *andon,* 464, 465–466
    hire method for using, 466
Painting process, reserved seat method example, 490
Paper lead-time, 4, 5
Parallel operations, 224–225, 536
    calculations for parts-production capacity work tables, 634

establishing in transfer machine blade replacement, 528
full *vs.* partial, 225
Pareto chart, 132, 457
Parking lots, well- and poorly-managed, 300
Parkinson's Law, 126
Part omission errors, 607
Partial parallel operations, 225
    calculations for parts-production capacity work tables, 633–634
Parts assembly
    preventing omission of parts tightening, 681
    preventing oversights with *jidoka*, 680–681
Parts development, 52
Parts inventories
    demand trends, 475
    strategies for reducing, 475–476
Parts list, depth and production method, 82
Parts placement
    in cooperative operations, 370
    standard operations improvements, 643
Parts tray/box, visible organization, 385
Parts waste, 157
Parts, improvements in picking up, 643–644
Parts-production capacity work table, 626, 629, 822–824
    serial operations calculations, 633
    steps in creating, 632–634
Pay raises, basis of, 394
PCB assembly plant, internal-external changeover improvements, 513–514
People
    as root of production, 104, 107, 108
    training for multi-process operations, 389
Per-day production total, 487
Per-unit time, 633
Performance below expectations, 686
Personnel costs, and manpower strategies, 63
Photoelectric switches, 572, 574, 682
    applications, 572
    object, detection method, and function, 573
Physical lead-time, 5
Pickup *kanban,* 444
Piecemeal approach, failure of, xiii
Pin dimension errors, 595
Pinch hitters, 407
Pitch, 66, 67, 337, 433, 469. *See also* Cycle time
    adjusting to worker pace, 358–359
    approaches to calculating, 485

factors determining, 70
failure to maintain, 678
hourly, 482
individual differences in, 67
myth of conveyor contribution to, 156
Pitch buzzers, waste prevention with, 232
Pitch per unit, 649
Plane improvements, 26–27
Plywood gluing process, accidents, 696
Pneumatic cylinders
    safety improvement from, 694
    workpiece removal with, 667
Pneumatic switches, 680–681
Point improvements, 25
    line improvements as accumulation of, 26
Poka-yoke, 119, 159, 177, 675, 680, 682. *See also* Safety
    and defect prevention, 566
    approaches, 568–570
    concept and methodology, 565–568
    control devices, 567
    defect prevention with, 564
    detection devices, 570–585
    drilling machine case study, 703
    for crane operations, 706
    for multi-process operations, 402–403
    milling machine example, 705–706
    safety applications, 703–709, 709
    safety cage on press, 704
    safety plate case, 703
    stop devices, 566–567
    warning devices, 567
Poka-yoke case studies, by defect type, 586–587
Poka-yoke checklists
    three-point evaluation, 619–620
    three-point response, 620–622
    using, 616–622
Poka-yoke detection devices, 570
    applications, 585
    contact devices, 570–572
    noncontact switches, 572–575
Poka-yoke/zero defects checklist, 820–821
Policy-based buying, 162
Position adjustments, avoiding, 537–538
Positioning sensors, 574
    applications, 577
Positive attitude, 204–205
Possible utilization rate, 684–685, 708
Postural ease, 221
Power, inexpensive types, 222
PQCDS approach, 2, 3

Practical line balancing, 357, 358
Preassembly processes, scheduling, 477
Preparation waste, in external changeover
      procedures, 510
Preset stop positions, 680
Press die alignment errors, 596
Press operator, waste example, 77–78
Presses
      safety problems, 702
      worker separation, 703
Preventive inspection, 564
Preventive maintenance, 688, 708
Previous process-dependent production, 54
Price cutting, due to inventory, 325
Printed circuit board assembly shop, 211
      arrow diagrams, 189, 212
Proactive improvement attitude, 54
Problem-solving, *vs.* evasive responses, 150
Process display standing signboards, 462–463
Process improvement models, 166, 167
Process route diagrams, 782–784
Process route tables, 347, 348
Process separation, 216, 219
Process waiting waste, 214, 218
Process, transfer, process, transfer system, 59
Process-and-go production, 55–59, 57, 59
Process-related waste, 177–178
Processing, 56, 160, 187
      lack of time spent in, 58
      shish-kabob *vs.* flow production
            approaches, 329–330
Processing errors, 586
Processing islands
      abolishment of, 396–398
      eliminating, 421, 426–427
Processing omissions, 586, 588–600
Processing sequence
      at diecast factory, 374, 376
      equipment layout by, 336–337, 353–355
Processing time, reducing to shorten
      lead-time, 55
Processing waste, 166–167, 180
Procrastination, 205, 207
Procurement
      applying JIT to, 47
      standardization, 229
Product inventory, 101, 102
      demand trends, 475
      strategies for reducing, 475–476
Product lead-time, 4
Product model changes
      and capacity utilization rates, 504

avoidance of, 162
Product set-up errors, 602
Product-out approach, 36, 416, 483, 555
      once-a-month production scheduling in,
            479
Product-specific delivery sites, 384
Product-specific inventory management, 305
Production
      equipment- *vs.* people-oriented, 112–113
      roots in people, 104, 108
      waste-free, 49
Production analysis, 345–348
Production as music, 29–50, 51–54
      three essential elements, 50
Production factor waste, 159–160
      conveyance and, 163–166
      inspection and, 167–172
      processing and, 166–167
      retention and, 160–163
Production input, 59, 60
Production *kanban,* 443, 445
Production leveling, 21, 421–422, 482.
            See also Leveling
      as prerequisite for efficiency, 71
      flow production development for, 492–494
      importance to efficiency, 69
      kaizen retooling for, 494–495
      strategies for realizing, 492–494
      with *kanban* systems, 442, 445
Production management
      conventional approach, 3–7
      defined, 6
      management system, 6
      physical system, 6
      vs. JIT production management, 1–3
Production management boards, 457,
            470–471, 802–804
Production method
      and defect prevention, 555
      shish-kabob *vs.* level production, 484, 486
Production output, 59, 60
      and in-process inventory, 89
      and volume of orders, 61
      increasing without intensifying labor, 86
Production philosophy, shish-kabob *vs.*
            level production, 483–484, 486
Production planning, 52
Production schedules, 4
      leveling production, 482
      once-a-day production, 480–482
      once-a-month production, 478–479

once-a-week production, 479–480
strategies for creating, 477
Production standards, 623. *See also* Standard operations
Production techniques, 715
JIT Management Diagnostic List, 718
Production work orders, *vs. kanban* systems, 437–439
Productivity, 59–61
and volume of orders, 61
boosting with safety measures, 701
coexisting with humanity, 387–388
volume-oriented approach to, 415
Productivity equation, 415, 416
Products, in PQCDS approach, 3
Profit
and cost reduction, 36
losses through motion, 77
Profitable factories, 40
anatomy of, 39
Progress *andon,* 464, 469–470
Proper arrangement, 16, 157, 243–245, 510
applied to die storage sites, 530–531
five-point checklist, 772
made visible through red tags and signboards, 265–268
obstacles to, 17
visible, 251–252
Proximity switches, 574
applications, 576
Pseudo improvements, 126–130
Public changeover timetable, 505, 808–809
Pull production, 10, 26, 51, 52, 54, 70, 438
flow of information and materials in, 53
relationship to goods, 439
using full work system, 367
using hand delivery, 366
vocal, 371, 372
Punching lathe, worker separation, 702
Purchasing-related *kanban,* 449–450
Push production, 10, 26, 51, 419, 438, 439
as obstacle to synchronization, 364–365
flow of information and materials in, 53

Q

QCD (quality, cost, delivery) approach, 2
Quality
estimated, 122

improving through 5Ss, 241
in PQCDS approach, 3
process-by-process, 123–126
Quality assurance, 724
and defect identification, 546–561
and *poka-yoke* system, 565–585
as starting point in building products, 541–542
in JIT production system, 11
JIT five levels of QA achievement, 542–546
poka-yoke defect case studies, 586–615
use of *poka-yoke* and zero defects checklists, 616–622
zero defects plan, 561–565
Quality check points, for standard operations charts, 636–638
Quality control inspection method, 169

R

Radar chart, 727
Rational production, 120–121, 122
shish-kabob *vs.* flow production approaches, 330
Reality-based leveling, 23
Recession-resistant production system, 8
Red tag campaign reports, 761–763
Red tag criteria, setting, 273–274
Red tag episodes, 281
employee involvement, 284
excess pallets, 283
red tag stickers, 283–284
red tagging people, 282
showing no mercy, 284–285
twenty years of inventory, 281–282
twice red tagged, 282
yellow tag flop, 283
Red tag forms, 271
Red tag items list, 765
Red tag list, computer-operated, 280
Red tag strategy, xii, 17, 265–268, 269–270, 455
campaign timing, 268
case study at Company S, 285–289
criteria setting, 273–274
for visual control, 268–269
implementation case study, 290–293
indicating where, what type, how many, 268

main tasks in, 291
making tags, 274–275
overall procedure, 267
project launch, 271, 273
red tag episodes, 281–285
red tagging computers, 278–281
steps, 270–278, 272
tag attachment, 276
target evaluation, 276–278
target identification, 273
understanding, 282
waste prevention with, 231
Red tag strategy checklist, 292
Red tag strategy report form, 293
Red tag targets
evaluating, 276–278
identifying, 273
Red tags, 758, 759, 760
attaching, 276
example, 275
making, 274–275
Reliability, increasing in equipment, 688
Reordering point method, 435–437, 475
Replacement waste, 509–510
eliminating in internal changeover,
518–523
Required volume planning, 52
Research and development, 37
Reserved carts, for changeover, 531–532
Reserved seat method, 489–490
painting process example, 490
Resistance, 42, 43, 199, 201–202
and arguments against JIT improvement,
200
and inventory accumulation, 322
by foremen and equipment operators, 30
from senior management, 15
to change, 41, 84
to multiple-skills training, 407
Responsiveness, 453
Retention, 56, 57, 160, 186, 187
and anticipatory buying, 162
and anticipatory manufacturing, 162
and capacity imbalances, 161–162
in shish-kabob production, 484
process, retention, transfer system, 59
reducing, 59
waste in, 160–163
Retention waste
eliminating, 213–214
lot waiting waste, 215–216
process waiting waste, 214

Retooling time, 633
Retooling volume, 633
Return on investment (ROI), inventory and,
95
Return to start position, 663
applying *jidoka* to, 666, 667
Returning waste, 511
Rhythmic motions, 221
Rules, for safety, 696, 697, 699

S

S-shaped manufacturing cells, 362
Safety, 152, 406, 725
basic training for, 698–699
defined, 698–699
for multi-process operations, 403–404
full-fledged, 70–709
in JIT production system, 12
in PQCDS approach, 3
in standard operations chart, 701
poka-yoke applications, 703–703
standard operations goals, 624
through 5Ss, 241
visual assurance, 707–708
Safety cage, 704
Safety check points, for standard operations
charts, 637
Safety improvement, pneumatic cylinders to
springs, 694
Safety plate, 703
Safety strategies for zero injuries/accidents,
699–709
Salad oil example, 312
Sales figures
and equipment improvements, 115
impact of seasons and climatic changes on,
97
Sales price, 36
basis in market price, 35
Sampling inspection, 120–122
Screw-fastening operation, waste in, 148
Searching waste, 154
Seasonal adjustments, 323–324
Seiketsu (cleanup), 16, 239, 246–247
Seiri (proper arrangement), 16, 238, 243–245
photo exhibit, 260
Seiso (cleanliness), 16, 239, 246

Seiton (orderliness), 16, 245–246, 328
  photo exhibit, 260
Self-inspection, 392
Senior management
  approval for 5S approach, 262
  ignorance of production principles, 88
  need to believe in JIT, 139
  on-site inspection by, 264
  responsibility for 5S strategy, 263
  role in awareness revolution, 14–15
  role in production system change, 3
  Seniority, as basis of pay raises, 394
Sensor assembly line, multi-process
      operations on, 363
Sequential mixed loads, 379
Serial operations, 224
  calculations for parts-production capacity
      work tables, 633
  eliminating, 527–529
Set-up
  applying human automation to, 669
  pre-manufacturing, 499
  unprocessed workpieces, 663, 667
Set-up errors, 560, 586, 601–606
Seven QC tools, 132, 133
Seven types of waste, 172–174
  conveyance waste, 176
  defect production waste, 176–177
  idle time waste, 178–179
  inventory waste, 175–176
  operation-related waste, 178
  overproduction waste, 174–175
  process-related waste, 177–178
Shared specifications, 419
Shish-kabob production, 10, 17, 18, 20, 46,
      70, 104, 166, 207
  approach to processing, 329–330
  as large-lot production, 423
  as obstacle to synchronization, 371–373
  disadvantages, 158
  equipment approach, 331
  equipment layout in, 330
  in-process inventory approach, 331
  lead time approach, 331
  operator approaches, 330–331
  production scheduling for, 476
  rational production approach in, 330
  vs. flow production, 328–332
  vs. level production, 482–485, 486
  waste in, 158
Shitsuke (discipline), 16, 239, 247–249
Short-delivery scheduling, 379, 497

Shotblaster
  at diecast factory, 375
  compact, 354, 377, 398–399
Shukan (custom), 689
Signal *kanban,* 443, 445, 446
Signboard strategy, 442, 455, 464
  amount indicators, 301–302
  and FIFO, 302–303
  defined, 294–296
  determining locations, 296
  die storage site using, 530
  for delivery site management, 383
  for visual orderliness, 293–294
  habitual orderliness, 302
  indicating item names, 299–300
  indicating locations, 298
  item indicators, 301
  location indicators, 299
  parking lot item indicator examples, 300
  preparing locations, 296–298
  procedure, 297
  signboard examples, 295
  steps, 296–302
Signboards, 43, 44, 265–268
  overall procedure, 267
  waste prevention with, 231
Simplified work procedures, 404
  and defect prevention, 549
Single-process workers, 339, 375, 419
Single-product factories, 71
Single-product load, 379
Sink cabinet factory, flow production
      example, 493
Skin-deep automation, 79
Slow-but-safe approach, 102–103
Small-volume production, xi, 2, 62, 278, 321,
      497
Social waste, 159
Solder printing process, flow of goods
      improvement, 641
Sorting inspection, 168, 169
Spacer blocks
  and manual positioning, 524–525
  eliminating need for manual dial
      positioning with, 526
Speaker cabinet processing operations,
      improvements, 646–647
Special-order production, 2
Specialization
  in Western *vs.* Japanese unions, 393–394
  vs. multi-process operations, 639

Specialized carts, for changeover operations, 532

Specialized lines, 371–373

Specialized machines, cost advantages, 332

Speed, *vs.* cycle time, 116

Spindle hole punch processing omission, 590

Spirit of improvement, 43, 44

Staff reduction, 62, 418

Standard operating processes (SOPs), 23

Standard operation forms, 626
  parts-production capacity work table, 626
  standard operations chart, 627–628, 628
  standard operations combination chart, 626, 627
  standard operations pointers chart, 626–627, 627
  steps in creating, 630–638
  work methods chart, 627

Standard operations, 24, 50, 65, 193–194, 224, 623, 708–709, 724
  and multi-skilled workers, 650–651
  and operation improvements, 638–649
  as endless process, 624
  combination charts for, 223–226
  communicating meaning of, 652
  cost goals, 624
  cycle time and, 625
  defined, 623
  delivery goals, 624
  eliminating walking waste, 645–649
  equipment improvements facilitating, 640
  equipment improvements to prevent defects, 640
  establishing, 628–630, 629–630, 654
  factory-wide establishment, 652
  forms, 626–628
  goals, 624
  implementing for zero injuries/accidents, 699–703
  improvement study groups for, 653
  improvements to flow of goods/materials, 638–640
  in JIT production system, 11–12
  materials flow improvements, 641
  motion waste elimination through, 639
  movement efficiency improvements, 642–643
  multi-process-operations improvements, 639
  need for, 623–624
  obtaining third-party help, 653

one-handed to two-handed task improvements, 644–645
  operational rules improvements, 639–640
  parts placement improvements, 643
  picking up parts improvements, 643–644
  preserving, 650–654
  quality goals, 624
  rejection of status quo in, 653
  reminder postings, 652
  role in JIT introduction, 23–24
  safety goals, 624, 697
  separating human work from machine work for, 640, 649–650
  sign postings, 652
  spiral of improvement, 629
  standard in-process inventory and, 625–626
  ten commandments for, 651–654
  three basic elements, 625–626
  transparent operations and, 628
  waste prevention through, 226
  wood products manufacturer's combination charts, 227
  work sequence and, 625
  workshop leader skills, 652, 653

Standard operations chart, 627, 628, 629, 631, 637
  safety points, 700, 701
  steps in creating, 630–632, 636–638

Standard operations combination chart, 193, 457, 626, 627, 629, 631, 825–826
  steps in creating, 634–636

Standard operations form, 831–833

Standard operations pointers chart, 626–627, 627

Standard operations summary table, 827–828

Standard parameters, changeover of, 499

Standardization
  of equipment, 421
  waste prevention by, 228–230

Standby-for-lot inventory, 161

Standby-for-processing inventory, 161

Standing signboards, 462–463

Standing while working, 19, 118, 355, 424, 425, 429
  and cooperative operations, 368
  as condition for flow production, 339
  in assembly lines, 355–359
  in multi-process operations, 399–400
  in processing lines, 359–360
  work table adjustments for, 360

Statistical inventory control methods, 475

Statistical method, 570
  poka-yoke, 659
Status quo
  denying, 205
  failure to ensure corporate survival, 15
  reluctance to change, 42
Steady-demand inventories, 476
Stockpiling, 160
Stop devices, 566–567
Stop-and-go production, 55–59, 57
Stopgap measures, 150
Storage, cutting tools, 318
Straight-line flow production, 340, 360
Subcontracting, applying JIT to, 47
Subcontractors, bullying of, 378
Sudden-demand inventories, 476
Suggestion systems, 36
Supplier *kanban,* 443, 444
Supplies management, 81
Surplus production, 323
  and defects, 549
Sweat workers, 74, 75
Symmetrical arm motions, 220–221
Synchronization, 363–364
  as condition for flow production, 337
  bottlenecked process obstacle, 364
  changeover difficulties, 373
  obstacles to, 364–368
  PCB assembly line, 366, 367
  push method as obstacle to, 364–365
  work procedure variations as obstacle to,
      367–371

T

Taboo phrases, 202
  Japanese watch manufacturer, 203
Takt time, 368, 469, 482
Tap processing errors, 606
Tapping operations, defect prevention,
      673–674
Temporary storage, 160
Three Ms, in standard operations, 623
Three Ps, 432
Three-station arrangements, 165
Time graph analysis, changeover
      improvements, 513
Time workers, 75
Tool bit exchange, one-touch, 522

Tool elimination
  as internal changeover improvement,
      519–520
  by transferring tool functions, 316
Tool preparation errors, 560, 587, 615
Tools
  5-point check for orderliness, 256
  applying orderliness to, 307
  close storage site, 311
  color-coded orderliness, 308–309
  combining, 314, 315
  easy-to-maintain orderliness for, 307
  eliminating through orderliness, 313–316
  indicators, 308, 309
  machine-specific, 311
  outlined orderliness, 309
Tools placement, 222
  order of use, 222
Top-down improvements, 134–139
Torque tightening errors, 599
Torso motion, minimizing, 221
Total quality control (TQC), 36, 132
Total value added, 715
Training
  for basic safety, 698–699
  for multi-process operations, 407–413
  for multiple skills, 400
  in CCO, 708
  in Japanese *vs.* Western factories, 395
Training costs, for multi-process operations,
      394–395
Transfer, 56, 57, 58
Transfer machine blade replacement, 528
Transparency, in multi-process operations,
      405
Transparent operations, and standard
      operations, 628
Transport *kanban,* 443
Transport routes, 380–382
Transportation lead-time, 99
Two-handed task improvements, 644–645
  and safety, 704
Two-process flow production lines, 360

U

U-shaped manufacturing cells, 340, 360–362
  as condition for flow production, 341
  for multi-process operations, 395–396

Unbalanced capacity, 322
Unbalanced inventory, 161, 322
Union leadership, 84
Unmanned processes, 668
Unneeded equipment list, 767
Unneeded inventory list, 765, 766
Unneeded items
    moving out, 266
    separating from needed items, 266
    throwing out, 266
    types and disposal treatments, 277
    unneeded equipment list, 278
    unneeded inventory items list, 277
Unprocessed workpieces, set-up, 663, 668
Unprofitable factories, anatomy of, 38
Usability testing, and defect prevention,
        549–550
Use points, maximum proximity, 222
Usefulness, and value-added, 147

V

Value analysis (VA), 157
Value engineering (VE), 157
Value-added work, 85, 166
    JIT Management Diagnostic List, 717
    vs. wasteful motion, 86, 147
VCR assembly line, cooperative operations
        example, 429
Vertical development, 20, 24–27, 26, 378, 391
Vertical improvement makers, 167
Vibration switches, 574
    applications, 583
Visible 5Ss, 249–251, 252
    visible cleanliness, 253
    visible discipline, 254–255
    visible orderliness, 252–253
    visible proper arrangement, 251–252
    visibly cleaned up, 253
Visible cleanliness, 253
Visible discipline, 254–255
Visible orderliness, 252–253
    with signboard strategy, 295
Visible proper arrangement, 251–252
Visibly cleaned up, 253
Visual control, 26, 120, 231, 251, 723
    and *kaizen,* 471–473
    andon for, 456, 464–470

as non-guarantee of improvements,
        453–454, 472–473
defect prevention with, 563
defective item displays for, 456, 457, 458
error prevention through, 456, 458
for safety, 700
in JIT production, 10–11
in *kanban* systems, 437
kaizen boards for, 462
kanban for, 456, 457
management flexibility through, 419
preventing communication errors with,
        556
process display standing signboards,
        462–463
production management boards for, 456,
        457, 470–471
red demarcators, 455, 456
red tag strategy for, 268–269, 455, 456
signboard strategy, 455, 456
standard operation charts for, 456, 457
standing signboards for, 462–463
through *kanban,* 440
types of, 455–459
visual orderliness case study, 459–462
waste prevention with, 230–232
white demarcators, 455, 456
Visual control tools, roles in improvement
        cycle, 473
Visual orderliness
    case study, 459–462
    in electronics parts storage area, 460
    signboard strategy for, 293–306
Visual proper arrangement, 17
Visual safety assurance, 707–708
Vocal pull production, 371, 372
Volume of orders, and production output, 61

W

Walking time, 635
Walking waste, 153–154, 173, 536
    eliminating for standard operations,
        645–649
Wall of fixed ideas, 210
Warehouse inventories, 161, 175
    as factory graveyards, 73
    reduction to zero, 20
Warehouse maintenance costs, 73

Warehouse waste, 69
Warning *andon,* 466–468
Warning devices, 567
Warning signals, 567
Washing unit, 364
   compact, 356
   in-line layout, 365
Waste, xii, 15, 643
   5MQS waste, 152–159
   and corresponding responses, 180
   and inventory, 48
   and motion, 75
   and red tag strategy, 269
   as everything but work, 182, 184, 191
   avoiding creation of, 226–236
   concealment by shish-kabob production,
      17, 158
   conveyance due to inventory, 90
   deeply embedded, 18, 150, 151
   defined, 146–150
   developing intuition for, 198
   eliminating with 5Ss, 508–511
   elimination by *kanban,* 440
   elimination through JIT production, xi, 8,
      341–342
   embedding and hiding, 84
   examples of motion as, 76
   hidden, 179
   hiding in conveyor flows, 67
   how to discover, 179–181, 179–198
   how to remove, 198–226
   identifying in changeover procedures,
      508–511
   in changeover procedures, 501
   in external changeover operations,
      510–511
   in internal changeover operations, 509–510
   in screw-fastening operation, 148
   inherited *vs.* inherent, 79–84
   invisible, 111
   JIT and cost reduction approach to, 69–71
   JIT Production System perspective, 152
   JIT seven types of, 172–179
   JITs seven types of, 172–179
   latent, 198
   making visible, 147
   minimizing through *kanban* systems, 437
   production factor waste, 159–172
   reasons behind, 146–150
   reinforcing by equipment improvements,
      111–112
   related to single large cleaning chamber,
      155
   removing, 84–86, 198–226
   severity levels, 171–172
   through computerization, 83
   total elimination of, 145, 152
   types of, 151–179
Waste checklists, 194–198
   five levels of magnitude, 195
   how to use, 195
   negative/positive statements, 197
   process-specific, 195, 196, 197, 198
   three magnitude levels, 197
   workshop-specific, 195
Waste concealment, 454
   by inventory, 326, 327
   revealing with one-piece flow, 350–351, 352
Waste discovery, 179–181
   back-door approach to, 181–183
   through current conditions analysis,
      185–198
   with arrow diagrams, 186–190
   with one-piece flow under current
      conditions, 183–185
   with operations analysis tables, 190–192
   with standard operations, 193–194
   with waste-finding checklists, 194–198
Waste prevention, 226, 228
   and do it now attitude, 236
   by avoiding fixed thinking, 235–236
   by outlining technique, 231
   by thorough standardization, 228–230
   with 5W1H sheet, 232–236
   with *andon,* 232
   with *kanban* system, 232
   with one-piece flow, 353
   with pitch and inspection buzzers, 232
   with red tagging, 231
   with signboards, 231
   with visual and auditory control, 230–232
Waste proliferation, 198, 199
Waste removal, 198–199
   50% implementation rate, 205–206
   and Basic Spirit principles for
      improvement, 204
   and denial of status quo, 205
   and eliminating fixed ideas, 204
   basic attitude for, 199–211
   by correcting mistakes, 207
   by cutting spending on improvements, 207
   by experiential wisdom, 210–211
   by Five Whys approach, 208–210

by using the brain, 208
in wasteful movement, 211–217
lot waiting waste, 219
positive attitude towards, 204–205
process waiting waste, 218
through combination charts for standard operations, 223–226
wasteful human movement, 217–223
Waste transformation, 198
Waste-finding checklists, 737–743
process-specific, 739, 741, 742, 743
workshop-specific, 738, 740
Waste-free production, 49
Waste-related forms, 730
5W1H checklists, 744–746
arrow diagrams, 730–732
general flow analysis charts, 733–734
operations analysis charts, 735–736
waste-finding checklists, 737–743
Wasteful movement
and eliminating retention waste, 213–217
by people, 217–223
eliminating, 211, 213
Wastology, 145
Watch stem processes, 397, 398
Watching waste, 154
Weekly JIT improvement report, 846–848
Whirligig beetle *(mizusumashi)*, 465
Wire harness molding process, internal changeover improvement case study, 517–518
Withdrawal *kanban*, 444
Wood products factory, multi-process operations in, 425
Work
as value-added functions, 182
meaning of, 74–75
motion and, 74–79
vs. motion, 657, 659
Work environment, comfort of, 223
Work methods chart, 627, 629, 829–830
Work operations, primacy over equipment improvements, 103–108
Work sequence, 636
and standard operations, 625
arranging equipment according to, 638
for standard operations charts, 636
Work tables, ergonomics, 222
Work-in-process, 8
management, 81, 83
Work-to-motion ratio, 86
Work/material accumulation waste, 173

Worker hour minimization, 62, 66–69
Worker mobility, 19
Worker variations, 367–371
Workerless automation, 106
Workpiece directional errors, 605
Workpiece extraction, 663
Workpiece feeding, applying automation to, 665
Workpiece motion, waste in, 158–159
Workpiece pile-ups, 25, 118
Workpiece positioning errors, 605
Workpiece processing, applying *jidoka* to, 664
Workpiece removal
applying human automation to, 668
motor-driven chain for, 695
with processed cylinders, 667
Wrong part errors, 587, 612, 613
Wrong workpiece, 560, 587, 614

Y

Yen appreciation, xi

Z

Zero accidents, 699
Zero breakdowns, 684, 685
production maintenance cycle for, 687
with 5S approach, 241
Zero changeovers, with 5S approach, 242
Zero complaints, with 5S approach, 242
Zero defects, 545
5S strategy for, 565
human errors and, 562–563
information strategies, 563
machine cause strategies, 564
material cause strategies, 564
overall plan for achieving, 561–565
production maintenance cycle for, 687
production method causes and strategies, 564–565
with 5S approach, 241
Zero defects checklists
three-point evaluation, 619–620

three-point response, 620–622
using, 616–622
Zero delays, with 5S approach, 242
Zero injuries
strategies for, 699–709

with 5S approach, 241
Zero inventory, 20, 98–102
importance of faith in, 176
Zero red ink, with 5S approach, 242
Zigzag motions, avoiding, 221

About the Author

Hiroyuki Hirano believes Just-In-Time (JIT) is a theory and technique to thoroughly eliminate waste. He also calls the manufacturing process the equivalent of making music. In Japan, South Korea, and Europe, Mr. Hirano has led the on-site rationalization improvement movement using JIT production methods. The companies Mr. Hirano has worked with include:

Polar Synthetic Chemical Kogyo Corporation
Matsushita Denko Corporation
Sunwave Kogyo Corporation
Olympic Corporation
Ube Kyosan Corporation
Fujitsu Corporation
Yasuda Kogyo Corporation
Sharp Corporation and associated industries
Nihon Denki Corporation and associated industries
Kimura Denki Manufacturing Corporation and associated industries
Fukuda ME Kogyo Corporation
Akazashina Manufacturing Corporation
Runeau Public Corporation (France)
Kumho (South Korea)
Samsung Electronics (South Korea)
Samsung Watch (South Korea)
Sani Electric (South Korea)

Mr. Hirano was born in Tokyo, Japan, in 1946. After graduating from Senshu University's School of Economics, Mr. Hirano worked with Japan's largest computer manufacturer in laying the conceptual groundwork for the country's first full-fledged production management system. Using his own interpretation of the JIT philosophy, which emphasizes "ideas and techniques for the complete elimination of waste," Mr. Hirano went on to help bring the JIT Production Revolution to dozens of companies, including Japanese companies as well as major firms abroad, such as a French automobile manufacturer and a Korean consumer electronics company.

The author's many publications in Japanese include: *Seeing Is Understanding: Just-In-Time Production* (*Me de mite wakaru jasuto in taimu seisanh hoshiki*), *Encyclopedia of Factory Rationalization* (*Kojo o gorika suru jiten*), *5S Comics* (*Manga 5S*), *Graffiti Guide to the JIT Factory Revolution* (*Gurafiti JIT kojo kakumei*), and a six-part video tape series entitled *JIT Production Revolution, Stages I and II*. All of these titles are available in Japanese from the publisher, Nikkan Kogyo Shimbun, Ltd. (Tokyo).

In 1989, Productivity Press made Mr. Hirano's *JIT Factory Revolution: A Pictorial Guide to Factory Design of the Future* available in English.

Printed and bound by CPI Group (UK) Ltd, Croydon, CR0 4YY

23/10/2024

01777685-0014